# Hazard Communication Made Easy

## A Checklist Approach to OSHA Compliance

Sean M. Nelson
John R. Grubbs, MBA, CIT

Government Institutes
Rockville, Maryland

Government Institutes, a Division of ABS Group Inc.
4 Research Place, Rockville, Maryland 20850, USA
Phone: (301) 921-2300
Fax: (301) 921-0373
Email: giinfo@govinst.com
Internet: http://www.govinst.com



04 03 02 01 00 5 4 3 2 1



Nelson, Sean M.
Hazard Communication Made Easy: A checklist approach to OSHA compliance /
Sean M. Nelson, John R. Grubbs.
p. cm.
Includes biographical references and index.

ISBN 0-86587-656-8

1. Industrial safety—Law and legislation—United States. 2. Hazardous substances—Law and legislation—United States.
I. Grubbs, John R. II. Title.

KF3570.N45 2000
344.73'0465—dc21

00-038954
CIP

Printed in the United States of America

*For*

*Traci, Mom, Dad, and Kim*

*and*

*Amie, Cameron, Mom, Dad, James, and Jerianne*

# Summary Contents

# Contents

# Figures and Tables

# Preface

In order to provide a useful tool for all professionals, we have developed *Hazard Communication Made Easy* to assist the reader in complying with OSHA's most cited regulation, Hazard Communication (HAZCOM), 29 CFR 1910.1200. This book breaks compliance into simple subparts including:

- An overview of the standard and its requirements with full explanations for compliance;
- Sample programs that can be easily manipulated for individual compliance; and
- Those chemicals that OSHA regulates specifically for industry in an easy to use alphabetical format.

The days of searching through endless volumes of regulations for HAZCOM requirements and compliance are over when the reader uses *Hazard Communication Made Easy*. With more and more emphasis on making the workplace safe, many professionals are overwhelmed by the enormity of regulatory requirements and the different interpretations of those regulations. Our focus is to provide an easy to use reference for HAZCOM compliance that will save the reader time and promote safety in the workplace while making informed, credible decisions.

Also, the complexity of the legal jargon in each regulation makes interpretation inconsistent and limits the understanding of what is necessary to develop a safe and healthy workplace. The explanations provided in this book will help you transfer the necessary information in a consistent and logical manner.

Based on our experience, we have provided you with our interpretation of the HAZCOM standard. If detailed compliance information and the regulations in their entire, unedited form is needed or desired, we recommend that you consult OSHA and the regulations directly. However, when you need quick information in a common sense, easy to use format, *Hazard Communication Made Easy* is the tool for you.

As you work through your regulatory compliance with *Hazard Communication Made Easy*, we encourage you to provide any feedback that can be used to improve this work. Remember, for you overall compliance needs, *Safety Made Easy, 2nd Edition*, is now updated and available from Government Institutes. Good luck and thank you for making *Hazard Communication Made Easy* a part of your organization's compliance effort.

Sean Nelson

John Grubbs

# About the Authors

**Sean Nelson** manages ground safety and hazardous materials operations for Atlantic Southeast Airlines, Inc. Responsible for maintenance facilities in Macon, Georgia; Atlanta, Georgia; Texarkana, Arkansas; and Dallas, Texas, his duties include safety and hazardous materials program development, auditing, consultation, training, and hazardous waste management. He also serves as an advisor for safety committees at each facility.

Sean is a formally trained safety specialist with a degree in occupational safety and health. His education includes instruction in OSHA law, hazardous materials management, hazardous waste management, risk management, worker's compensation laws, and industrial hygiene. Sean is also certified as a trainer for shipping hazardous materials and waste. In addition, he holds active memberships with the American Society of Safety Engineers, Middle Georgia Safety Council, and the National Association of Industrial Technology.

**John Grubbs, MBA, CIT,** is currently managing a special project at Lone Star Steel Company in Lone Star, Texas. Prior to this assignment, John was a safety engineer with duties that included safety training, safety audits, safety engineering and design, industrial hygiene sampling, ergonomic studies, and work toward attaining OSHA VPP "STAR" status. He holds undergraduate degrees in occupational safety and health and industrial technology as well as a Masters in Business Administration.

Prior to working for Lone Star Steel, John was the corporate safety coordinator for National Linen Service in Atlanta, Georgia where he conducted industrial and fleet safety training for 10,000 employees and 3,000 DOT drivers at over 100 facilities. John also has safety experience in the ceramic tile manufacturing industry. He is a Certified Industrial Technologist and is registered with the Texas Department of Health as an Emergency Care Attendant. He is currently the President of the Arklatex Chapter of the American Industrial Hygiene Association and a member of the American Society of Safety Engineers and National Association of Industrial Technology.

Part I

# Introduction to Hazard Communication

## Background

The Occupational Safety and Health Administration established the Hazard Communication Standard (HCS) in 1983. Over the years the HCS has been revised to include all employees. The HCS or "Right-to-Know" standard was established to ensure that all employers comply with the requirements of the standard and to inform employees about the hazardous substances with which they work.

The standard covers all employees working with hazardous substances in all forms including open or closed containers, air contaminants, piping systems, and storage tanks. Hazardous substances can be defined as any material that contains a physical or health hazard. This includes any material in the form of a solid, liquid, gas, vapor, fume, or mist that may be hazardous because it is flammable, explosive, corrosive, toxic, reactive, or an oxidizer.

By complying with this standard, employees can learn the hazards of the products for which they work, methods for protection, and receive a visual alert of the presence of a hazardous substance. The employer has the advantage of reduced liability, accident, and illness rates, and obtaining knowledge for reducing personnel and environmental exposure.

## Regulatory Responsibilities

Regulatory responsibilities will differ between companies depending on their size, type of business, and the state of business operations. There are many factors to consider when assigning responsibilities, developing a customized program, implementing a program, and training. Some of these include the number of products used, the method for use, potential exposure of employees to hazardous substances, and environmental and community impact. The good thing about the Hazard Communication Standard, is that OSHA has developed the standard to be performance-oriented. This means that you have the flexibility to adapt the rule to the needs of your workplace, rather than having to follow specific, rigid requirements. However, it also means that you have to exercise more judgment to produce an appropriate and effective program. In addition, the standard exempts from training, any office employees who only encounter hazardous chemicals in isolated and non-routine cases.

The size of the company can often determine the number of products that are used within the facility. A large company may use a diverse number of products in several different shops throughout the facility. Or, the entire company may use only a limited amount of products. A multi-facility company may use several products at each facility. One program coordinator must organize all this information. However, in large facilities or multi-facility companies, you may want to assign the hazardous materials manager to assist with maintenance of the program.

The type of business is a concern. Chemical manufacturers and importers are required to evaluate the hazards of the chemicals they produce or import. Using this information, they must prepare labels for containers and material safety data sheets (MSDSs). Chemical manufacturers, importers, and distributors of hazardous chemicals are all required to provide the appropriate labels and material safety data sheets to the companies to which they ship the chemicals. Companies can rely on the information received from suppliers and have no independent duty to analyze the chemical or evaluate the hazards of it.

The state in which the facility operates is also a concern. There are several states known as "State Plan States". This means that the state has its own hazard communication or "right-to-know" laws. Since federal law has jurisdiction over state law, these state plans must be at least as stringent as the federal standard. Twenty-three of the 50 United States are State Plan States. However, many of these states have not made changes to the federal HCS. The following states and territories have approved state plans:

Alaska
Arizona
California
Connecticut*
Hawaii
Indiana
Iowa
Kentucky
Maryland
Michigan
Minnesota
Nevada
New Mexico
New York*
North Carolina
Oregon
Puerto Rico
South Carolina
Tennessee
Utah
Vermont
Virgin Islands
Virginia
Washington
Wyoming

*The Connecticut and New York plans cover public sector (state and local government) employment only.

OSHA's website provides links to more information concerning the State Plan States (http://www.osha-slc.gov/fso/osp/). Visit your state's site for more information, or contact your regional OSHA office listed in Appendix B of this text.

Lastly, remember to emphasize that everyone is responsible for compliance with this program. Both management and employees have specific responsibilities for assuring success and avoiding heavy penalties, not to mention a reduction of illness and injury occurrence.

## About Your HazCom Program

The sample program in this book has been provided as a tool to assist you with compliance of the Hazard Communication Standard. It contains the basic information needed to get most companies started with their compliance effort. It is generally all the information some companies need. However, others will need to further customize their program to the specific hazards within the facility. For example, a spray finishing facility should include a section that discusses the specific hazards of spray finishing. Your facility may contain one of the hazards mentioned in Section IV of this manual. Complete an overall assessment of your facility prior to program implementation to discover any hazards that may have otherwise been overlooked.

An accurate inventory of all hazardous substances within your facility is also very important. This inventory makes up your Workplace Chemical List. A Workplace Chemical List form is provided in Appendix A. The items from the Workplace Chemical List must be evaluated to determine the hazards for which you should train. If the evaluation shows that the products at your facility only contain flammable and corrosive hazards, then train specifically for those hazards. However, if a toxic hazard is introduced to the workplace, you must retrain for toxic issues.

Remember, this program must be continuously maintained. If your facility receives few new products, revise the Workplace Chemical List each time. If several items are received each month, set a schedule for revision. Make sure each copy is accurately revised and made available to employees of each shift at all times.

## About Specific Requirements for Chemical and Physical Hazards

Part IV in this book give the requirements for individual chemical and physical hazards such as lead, formaldehyde, and asbestos. Although these items have been broken down to remove the "loops" from the regulations, you must still have a working knowledge of the hazard to understand the particular section. Because of the technical nature of these items, there is really no way to provide the information in a more simple way without changing the integrity of the regulation itself.

# Your Company's Hazard Communication Program

## Getting Started

The goal of the sample program on page 9 is to provide an easy-to-complete solution to your compliance with the Hazard Communication Standard (HCS). There are a few elements that must be provided by your company to make the program complete. These elements are explained below. (A compliance checklist is also provided.)

- ❑ **Appoint a Program Coordinator.** The Program Coordinator is the person with overall responsibility for the program. This includes the completion of the written program, the implementation process, and maintenance of the program. (*Maintenance* of the program is not meant to suggest that the Program Coordinator is necessarily responsible for revisions, training, etc. of the program, but is responsible for naming the responsible individuals, and assuring their compliance with the standard through audit or observation.)

- ❑ **Create an inventory of hazardous materials throughout the facility.** For an initial assessment of materials found throughout the facility, it is a good idea for the Program Coordinator to assign a responsible person (manager/supervisor/lead) from each department to complete an inventory of materials. An inventory sheet should be very specific for the information you seek. *A sample inventory sheet may be found in Appendix A.*

- ❑ **Gather MSDSs for all hazardous materials.** Request an MSDS from the manufacturer of each product for which an MSDS cannot be found or for which the MSDS is more than three years old.

- ❑ **Complete the written program with the required information.** Most elements of the program have been provided for you. There are a few blanks in the sample program that should be filled in to complete the written requirement of the HazCom Standard. Also, a workplace chemical list should be completed using the inventory of chemicals previously requested. *A sample workplace chemical list can be found in Appendix A.* Read each of the endnotes listed behind the program. These endnotes contain important information or variations for your program.

- ❑ **Label all products throughout the facility** – Label each product as specified in your HazCom Program. The actual labels may be purchase from several different vendor choices. You should always be consistent with the type of label you choose (Same style of HMIS label). In larger companies or companies that use large quantities of a single product, you may consider purchasing a computerized labeling system or pre-printed labels.

- ❑ **Determine training responsibilities** – As appropriate, determine the method for which your company should complete training requirements. For instance, a small company may have department managers or supervisors administer a module-based training and maintain records within their department or forward them to the Human Resource Department that maintains personnel records. A larger company may have a separate training department that can issue the same module-based training or hold an interactive class style of training. Determine the number of employees that will require training and the level of training that is appropriate for each position (awareness level for warehouse employees, etc.) If training in a class setting, schedule according to level of training and convenience of the work shift (Train night shift employees immediately after their shift is complete. You don't want to train in the middle of the afternoon and expect the same employees to return that evening.). Also, schedule small classes. Larger classes tend to be impersonal and often cause the "confused individual" to be overlooked.

- ❑ **Provide appropriate training to personnel** – Warehouse personnel are only required limited HazComm training. Other personnel training must be more function-specific. For example, if a process within the company contains underlying hazard exposures to chemical material (tank cars filled with a toxic liquid), the employee must be informed of its hazards, additional safety precautions that may be required (confined space entry procedures), and emergency response information. Make sure all materials are prepared ahead of time and the materials are completely understood. Tips for effective training are given in Part III of this manual. We have also included a training module for your convenience.

- ❑ **Recordkeeping** – The standard does not require employers to keep records of employee training. OSHA compliance officers will talk to employees to determine if they have received training. Good management practice, though, is to maintain records on all training and the sample program in this manual has included procedures to do so. Maintain accurate records of all training and revisions to your HazCom Program. The Program Coordinator or appointed responsible person (local manager) must track all training to ensure accuracy and that all records are complete and current ***(NOTE: The standard does not require recurrent training for employees. This practice is to the discretion of the employer. Remember that it is the employer's ultimate responsibility for ensuring that employees are adequately trained.)***. The person responsible for records should keep them in a designated location that is consistent throughout the company (e.g. in the Training Department files at a central location or in a local department file where all other personnel records may be kept.). If applicable, this person should also issue a notice of required training for those individuals that are due recurrent training.

- ❑ **Copy of the Standard** – Include a copy of the standard at the end of your program for employee reference. A copy has been included at the end of the sample program and may be excerpted for this purpose.

# Checklist for Compliance

The following is a checklist for compliance that includes the advisory portion of the HCS published by OSHA. The sample program contains specific requirements and additional information concerning each subject.

- ❑ **Obtain a copy of the rule.** A copy of the HCS has been provided at the end of the sample program.

- ❑ **Assign responsibility for tasks.** Blanks have been provided throughout the sample program where specific personnel have designated responsibilities (A list of required entries may be found on the following page.) Other responsible positions or departments have been incorporated into the program.

- ❑ **Prepare an inventory of all chemical products.** A Chemical Inventory List Form is included in Appendix A.

- ❑ **Ensure containers are labeled.** This should be accomplished during implementation of the program and maintained by the appointed responsible person(s). The following list provides label requirements:

  - Chemical manufacturers and importers shall ensure that each chemical is properly labeled with its identity, all appropriate hazard warnings, and the name and address of the manufacturer.
  - The employer shall ensure that all chemicals in the workplace are properly labeled with the identity and hazard warnings as determined by the employer's policy or Hazardous Material Identification System (HMIS).
  - Portable containers do not require labels if used for a single shift by one employee.
  - Labels shall be legible and in English.

- ❑ **Obtain an MSDS for each chemical.** An MSDS Request Form is provided in Appendix A. Each MSDS shall include the following minimum information (see the sample MSDS form contained in the sample program):

  - common name and chemical name
  - manufacturer information
  - hazardous ingredients
  - health hazard data
  - physical data
  - fire and explosive data
  - reactivity data
  - spill or leak procedures
  - personal protection information
  - special precautions
  - SARA Title III information.

- **Prepare written program.** Develop your own program or use and revise the sample program provided in this manual.

- **Make MSDSs available to workers.** A good idea for accomplishing this task, depending on the number of MSDSs, is to include the MSDSs in a binder behind the HazCom program and other attachments. If your establishment has 250 or more products, you may want to consider a contract or MSDS-on-demand service.

- **Conduct training of workers.** Included with this manual (Part III) is a training module that you may use or revise to meet your requirements. A Training Certification Form is provided in Appendix A.

- **Establish procedures to maintain current program.** This is included in the sample program.

- **Establish procedures to evaluate effectiveness (of program).** This is included in the sample program.

## Program Information to be Completed by Company

Complete the following information to adapt the sample program to your company:

- Information Page Entries (company name, address, name of the Program Coordinator, and emergency phone numbers).

- Locations of the HazCom Program Manuals.

- Person, position, or department responsible for ensuring the training responsibilities of the program are met.

- Person, position (recommended), or department responsible for determining and affixing proper labels to products that are received.

- Position or department from which local or department managers may request training packages.

- Position or department where the "Right-to-Know" posters may be requested. Or, these may be purchased from a local safety supply company.

- Any additional forms (Workplace Chemical List, MSDS Request Form, Training Certification Form) needed to complete the program. (See Appendix A.)

> The sample Hazard Communication Program (pages 9-52) may be excerpted in its entirety to serve as a template for your company's Hazard Communication Program. You may also obtain an electronic version of this template by calling Government Institutes at 301-921-2355.

# Hazard Communication Program

## Information Page

Company Name: ______________________________

(The above company name, "Company", and "The Company" are terms that may be used interchangeably throughout this program.)

Facility Address/Location: ______________________________

______________________________

Program Coordinator: ______________________________

Emergency Phone Numbers:

Fire: ____________________

Medical: ____________________

Spill Response: ____________________

Revisions:

| Entered by/Date | Entered by/Date |
|---|---|
| 1.________/________ | 11.________/________ |
| 2.________/________ | 12.________/________ |
| 3.________/________ | 13.________/________ |
| 4.________/________ | 14.________/________ |
| 5.________/________ | 15.________/________ |
| 6.________/________ | 16.________/________ |
| 7.________/________ | 17.________/________ |
| 8.________/________ | 18.________/________ |
| 9.________/________ | 19.________/________ |
| 10.________/________ | 20.________/________ |

# Contents



1. **PURPOSE**: The purpose of this program is to ensure that ***The Company*** complies with the OSHA Hazard Communication Standard (HCS) as defined by Title 29 CFR Part 1910.1200. This program will provide information for the use of MSDSs, chemical product labeling, handling and storage, training, documentation, and recordkeeping requirements.

2. **GENERAL**: The Hazard Communication (HazCom) Program was designed to ensure that all employees and contracted personnel are made aware of and specifically trained in, the hazards associated with any hazardous chemical they may be exposed to while working on any company leased or owned facility. This program is applicable to all employees that handle, label, store, use, or have the potential to come in contact with chemical products at any time.

The Hazard Communication Program manuals, which contain all MSDS files, will be readily accessible to all employees during each work shift and are located at the following locations:

- **Master HazCom Program Manual** – Office of Program Coordinator
- ______________________________
- ______________________________
- ______________________________
- ______________________________

## 3. RESPONSIBILITIES:

A. It is Management's responsibility to assign a Program Coordinator to administer and maintain the HazCom Program.

B. The Program Coordinator is responsible for ensuring this program is maintained current and for evaluating it on a periodic basis to assure its effectiveness.

C. ______________________________ is responsible for ensuring that the training requirements of this program are met.

D. The facility supervisors are responsible for assuring compliance with this program.

E. ______________________________ is responsible for labeling containers that are brought into the facility.

F. Personnel trained and certified to ship hazardous materials will be responsible for labeling containers that are prepared for shipment.

G. Each affected employee is responsible for reading, understanding, and complying with this program. Any questioned responsibilities should be addressed by the Program Coordinator upon being informed of any conflict with the wording of this program or individual case misunderstanding.

H. An employee or department requiring or requesting a product that has not yet been used is responsible for submitting a written request to the Program Coordinator or assigned chemical advisory committee for approval of the product.

**4. TRAINING REQUIREMENTS**: All personnel who are or may be exposed to hazardous and/or toxic chemicals while performing duties in their work area must receive training on these hazards at the time of their initial assignment and whenever a new hazard is introduced into their work area. Persons who have not received this training will not be allowed to work in areas where hazardous chemicals or substances exist. Personnel must receive recurrent training within 12 months of initial training and at 12-month intervals thereafter.

Each employee will be informed of the following:

A. The requirements of the Hazard Communication Program and OSHA regulations;

B. Any operation in the employee's work area where hazardous chemicals are present; and

C. The location and availability of the written HazCom Program, including the required list(s) of hazardous chemicals and MSDSs.

**5. TRAINING CONTENT**: Content will include the following:

A. Methods to observe the work area to detect the presence or release of a hazardous chemical by visual appearance or the odor of the chemical when released;

B. Training on the physical and health hazards of the chemicals in their work area;

C. Training on the "Right-to-Know" of hazards (how to read and understand MSDSs);

*Note:* *Personnel without complete comprehension of a chemical's hazards or its MSDS should consult a supervisor. DO NOT PROCEED WITHOUT THIS KNOWLEDGE!*

D. An explanation of ***The Company***'s labeling system also known as the Hazardous Materials Identification System (HMIS);

E. Training media and answer keys for local management and/or designated individuals who will be responsible for administering the training (Training packages may be requested from ________________________________. Specify the number of packages required); and,

F. A quiz . The employee's supervisor should review any incorrect answers from the quiz to assure the employee has an understanding of the material and its application.

*Note:* *Local management will administer the training package. Upon review of the package, each student must complete the quiz and return to his/her supervisor for correction and discussion. This training must be documented on the appropriate form indicating that the student understands the program and any missed questions. The most recent quiz will be kept on local file for one year. If applicable, local management is responsible for assuring that recurrent training is completed as required.*

6. **OUTSIDE CONTRACTOR NOTIFICATION**: To ensure that any outside contractors work safely in any ***Company***-owned or operated facility, the local manager/supervisor will be responsible for providing the contractor with the following information:

A. Copies of the Hazard Communication Program along with a copy of the Workplace Chemical List;

B. The location of the MSDS files; and,

C. Information and location for any hazardous substances to which they may be exposed while working on the job site.

Contractors are required to review this information with their employees prior to the start of work at any ***Company*** facility. It will be the responsibility of the contractor to notify management of any hazardous chemicals that may be brought onto the job site.

7. **MATERIAL SAFETY DATA SHEETS (MSDSs)**: A Material Safety Data Sheet is a document which describes the physical and chemical properties of products, their physical and health hazards, and precautions for safe handling and use.

The manufacturers and distributors of chemical products furnish MSDSs. An MSDS must be maintained for each chemical used at the location. All departments should coordinate activities to ensure that all incoming shipments of hazardous substances arrive with an MSDS. Although physical attachment of the MSDS is not required on a shipment, it should accompany or precede the shipment.

Should a new product be received without an MSDS, DO NOT USE THE PRODUCT. It will be the responsibility of the requesting department manager or purchasing agent to request an MSDS for the particular product using an MSDS Request Form. Without the MSDS, you can not be certain of its health hazards or required personal protective equipment. If, after a reasonable amount of time, a response to the request is not received, a second letter requesting either an MSDS or disclaimer that the material is not hazardous will be sent.

*Note:* *Copies of the request forms should be made and placed with the local MSDS file. These copies will document the fact that an attempt was made to determine the hazards of the product.*

If, after the second request, no response is received, discontinue holding the product and send it back. Seek a replacement with another vendor.

In cases where a department receives repetitive shipments of the same chemical from a supplier, the MSDS may already be on file. Check the revision date and the date of the MSDS. If they are the same, discard the copy. If more recent, send to the Program Coordinator. This is to identify those situations where a "new" hazard associated with an existing chemical has been identified, or a new ingredient has been included in a currently used product.

**All new or revised MSDSs must be marked indicating the location for use of the product, and copied and forwarded to the Program Coordinator. Once the MSDS has been received and reviewed, it will be assigned a number for reference and sent back. This numbered MSDS should be inserted into the local MSDS manual, as appropriate, replacing the previous.**

The HazCom Standard does not describe the format to be used in completing the MSDS. However, the standard does specify the information the MSDS must contain (Title 29 CFR 1910.1200(g)). OSHA Form 174 meets the HazCom Standard's requirements and is recommended (See Figure 2-1).

| | | |
|---|---|---|
| | **Material Safety Data Sheet**<br>May be used to comply with<br>OSHA's Hazard Communication Standard,<br>29 CFR 1910.1200. Standard must be<br>consulted for specific requirements. | **U.S. Department of Labor**<br>Occupational Safety and Health<br>(Non-Mandatory Form)<br>Form Approved<br>OMB No. 1218-0072 |
| A | IDENTITY (As Used on Label and List) | Note: Blank spaces are not permitted. If any is not applicable, or no information is available, the space must be marked to indicate that. |
| | **Section I** | |
| B | Manufacturer's Name | Emergency Telephone Number |
| | Address (Number, Street, City, State, and ZIP Code) | Telephone Number for Information |
| | | Date Prepared |
| | | Signature of Preparer |

C **Section II --- Hazardous Ingredients/Identity Information**

| Hazardous Components (Specific Chemical Identity: Common Name(s)) | OSHA PEL | ACGIH TLV | Other Limits Recommended | % (optional) |
|---|---|---|---|---|
| | | | | |
| | | | | |
| | | | | |
| | | | | |
| | | | | |
| | | | | |
| | | | | |
| | | | | |
| | | | | |
| | | | | |
| | | | | |

D **Section III --- Physical/Chemical Characteristics**

| | | | |
|---|---|---|---|
| Boiling Point | | Specific Gravity (H2O = 1) | |
| Vapor Pressure (mm Hg.) | | Melting Point | |
| Vapor Density (AIR = 1) | | Evaporation Rate<br>(Butyl Acetate = 1) | |
| Solubility in Water | | | |
| Appearance and Odor (E) | | | |

**Section IV --- Fire and Explosion Hazard Data**

| | | | | |
|---|---|---|---|---|
| F | Flash Point (Method Used) | Flammable Limits G | LEL H | UEL |
| | Extinguishing Media | | | |
| I | Special Fire Fighting Procedures | | | |
| | | | | |
| | Unusual Fire and Explosion Hazards | | | |
| | | | | |
| | (Reproduce locally) | | | OSHA 174, Sept. 1985 |

**Figure 2-1: Sample MSDS (Front)**

J

| Section V --- Reactivity Data | | | |
|---|---|---|---|
| Stability | Unstable | | Conditions to Avoid |
| | Stable | | |

K

| Incompatibility (Materials to Avoid) | | | |
|---|---|---|---|
| Hazardous Decomposition or Byproducts | | | |
| Hazardous Polymerization | May Occur | | Conditions to Avoid |
| | Will Not Occur | | |

L

| Section VI --- Health Hazard Data | | | |
|---|---|---|---|
| Route(s) of Entry: | Inhalation? | Skin? | Ingestion? |

M

| Health Hazards (Acute and Chronic) | | | |
|---|---|---|---|
| | | | |
| | | | |

N

| Carcinogenicity: | NTP? | IARC Monographs? | OSHA Regulated? |
|---|---|---|---|
| | | | |

O

| Signs and Symptoms of Exposure |
|---|
| |

P

| Medical Conditions Generally Aggrivated by Exposure |
|---|
| |

Q

| Emergency and First Aid Procedures |
|---|
| |

R

| Section VII --- Precautions for Safe Handling and Use |
|---|

S

| Steps to Be Taken in Case Material Is Released or Spilled |
|---|
| |
| |

T

| Waste Disposal Method |
|---|
| |

U

| Precautions to Be Taken in Handling and Storing |
|---|
| |
| Other Precautions |
| |

V

| Section VIII --- Control Measures | | |
|---|---|---|

W

| Respiratory Protection (Specify Type) | | |
|---|---|---|
| Ventilation | Local Exhaust | Special |
| | Mechanical (General) | Other |

X

| Protective Gloves | Eye Protection |
|---|---|
| Other Protective Clothing or Equipment | |
| Work/Hygenic Practices | |
| | |

**Figure 2-1 (cont.): Sample MSDS (Back)**

**8. MSDS (OSHA FORM 174) REFERENCE EXPLANATION:**

A. Section I – Chemical Identification – The identity of the material must indicate specific chemical name, common name, and trade name, if applicable. The specific chemical name is one that will permit any qualified chemist to identify the structure of the compound(s) present to enable further searching of literature for desired information.

B. Name and address of the chemical manufacturer. Phone number to be used for information or an emergency.

C. Section II – Hazardous Ingredients/Identity Information – Lists content of the substance that may cause harm. The chemical name(s) of the substance, as well as the common name of the substance must be indicated. It also lists the concentration of the chemicals to which you can safely be exposed to within and eight-hour period (PEL – permissible exposure limit, or TLV – threshold limit value).

D. Section III – Physical/Chemical Characteristics – Describes the substance's appearance, odor, and other chemical characteristics. These are things that can affect the degree of hazard workers face in different work situations.

- Boiling Point/Melting Point – to help prevent a potentially dangerous change in state, as from a liquid to a breathable gas.
- Vapor Pressure, Density, and Evaporation Rate – important for flammable or toxic gases and vapors.
- Solubility in Water and Specific Gravity – identifies whether a chemical will dissolve in water, sink, or float.

E. Normal Appearance and Odor – Helps you recognize anything different and possibly dangerous.

F. Section IV – Flash Point – Indicates the minimum temperature at which a liquid gives off a vapor in sufficient concentration to ignite.

G. Flammable Limits – Indicates the concentration of the substance, in the form of a gas or a vapor that is required for it to ignite.

H. LEL and UEL – Lowest and highest (respectively) concentration (percentage of substance in the air) that will produce a flash of fire when an ignition source is present. At higher concentrations, the mixture is too "rich" to burn.

I. This section describes methods of firefighting including the proper extinguishing media (i.e. ABC, $CO_2$, foam, etc.). This section also describes unusual fire and explosion hazards.

J. Section V – Reactivity Data – Indicates whether the substance is stable or unstable and conditions to avoid, such as direct sunlight, or any condition which may cause a dangerous reaction.

K. Incompatibility – Lists materials, chemicals, and other substances to avoid which may cause the chemical to burn, explode, or release dangerous gases.

L. Section VI – Health Hazard Data – This is probably the most important section of the MSDS. It describes how a substance can enter your body, for instance:

- Inhalation (respired)
- Through the skin (absorbed)
- Swallowing (consumed)

M. Health Hazards (Acute and Chronic) – Lists the specific possible health hazards which could result from exposure, both acute and chronic.

Some effects like minor skin burns, are acute (they show up immediately after exposure).

Others, like liver damage, are chronic (they are often the result of exposure long ago or repeated small exposures over a long period of time).

N. If the substance or chemical is considered or believed to be a carcinogen, it will be indicated here.

O. This section lists the signs and symptoms of overexposure such as eye irritation, nausea, dizziness, skin rashes, headache, and burns.

P. If exposure to the substance could aggravate an existing medical condition, this information will be indicated here.

Q. This section describes emergency and first aid procedures to be followed in case of overexposure until medical help arrives. If these instructions are unclear, seek medical assistance immediately.

R. Section VII – Precautions for Safe Handling and Use – Explains procedures for spills, leaks, or any release; how to handle and store the substance; how to safely dispose of the substance; and other precautions.

S. This section describes the proper procedures to be followed in the case of spills or leaks, and the equipment needed.

*Note:* *Not all MSDSs will indicate the personal protective equipment recommended when cleaning up a spill. Always refer to section VIII – Control Measures, for specific types of recommended personal protective equipment to be worn prior to cleaning up a spill.*

Always notify your supervisor when a chemical spill or leak occurs, no matter what chemical, no matter how small.

T. This section lists the proper disposal procedures.

U. This section describes the proper handling and storage procedures, as well as other precautions to be followed to protect you and others (e.g. grounding containers during transfer of flammable materials to prevent static electricity as an ignition source).

V. Section VIII – Control Measures – Lists control measures for preventing or reducing the chance of harmful exposure, including personal protective equipment to be worn, work and hygiene practices, and ventilation requirements.

W. This section includes proper respiratory equipment and ventilation information.

X. This section describes gloves, eye protection, and other personal protective clothing recommended or required.

This section lists work or hygiene practices which should be exercised when handling this substance or chemical such as:

- Taking a shower
- Washing work clothes
- Destroying soiled clothing
- Washing hands

**9. WORKPLACE CHEMICAL LIST**:

A. General – The Workplace Chemical List identifies stock materials that are or contain ingredients classified by OSHA and/or the DOT as potentially hazardous to personnel. The Hazard Communication Standard requires that a list of hazardous chemicals be included as part of the company's Hazard Communication Program. The list will serve as an inventory for which an MSDS is maintained on file. Supplemental information contained on the Workplace Chemical List is intended only for the convenience or quick reference of the employee.

Workplace Chemical Lists are located behind the Hazard Communication Program. The Program Coordinator prepares the list with information compiled from the respective MSDS supplied by the manufacturer of the hazardous material.

B. Sample – Figure 2-2 shows a sample Workplace Chemical List, Section C provides a detailed explanation of the rows on the list. This list will give information that will assist in the safe handling and proper labeling of materials. The lists in the workplace are preceded by an index identifying the letters assigned to the specified rows (these may be found in Tables 2-1 through 2-5).

| 1 | 2 | 3 | 4 | 5 | 6 | 7 | 8 | 9 | 10 |
|---|---|---|---|---|---|---|---|---|---|
| **MSDS#** | **Description** | **MSDS Date** | **Health Hazard** | **PPE** | **Flashpoint** | **First Aid** | **Spill** | **Disposal** | **Location** |
| 1000 | Acetone, ABC Chemical Corp. | August 1998 | Health – 1<br>Fire – 3<br>Reactivity Specific – 0 | B | 5°F | Eyes – A<br>Skin – A<br>Inhale – A<br>Ingest – A | B | A | Paint Shop |
| 1120 | Lacquer Thinner, Best Paint Company | December 1998 | Health – 2<br>Fire – 3<br>Reactivity Specific – 0 | H | 21°F | Eyes – A<br>Skin – A<br>Inhale – A<br>Ingest – A | A | A | Paint Shop |
| 1230 | Methyl Ethyl Ketone | April 1998 | Health – 3<br>Fire – 3<br>Reactivity Specific – 0 | I | 23°F | Eyes – A<br>Skin – A<br>Inhale – A<br>Ingest – A | B | A | All |

**Figure 2-2: Sample Workplace Chemical List**

C. Column Explanations

1. MSDS Number – Identifies the product listed in the second column with an MSDS number. This will serve as the product's permanent reference number.

2. Description – The name of the chemical and, in some cases, a brief description of the product.

3. MSDS Date – The date the MSDS was produced.

4. Health Hazard Rating – The health hazard rating (NFPA rating system) for the product listed in the second column. This number rating is defined in Table 2-1.

   Fire Hazard Rating – The flammable rating (NFPA rating system) for the product listed in the second column. This number rating is defined in Table 2-1.

   Reactivity Hazard Rating – The reactivity hazard rating (NFPA rating system) for the product listed in the second column. This number rating is defined in Table 2-1.

5. PPE – A letter reference as to the personal protective equipment recommended for working with the product listed in the second column. See Table 2-2 for a definition of the referenced letters.

6. Flashpoint – Refers to the temperature, usually in degrees *F*, at which the total product listed in the second column, is capable of producing enough flammable vapors to ignite. Products having a flash point below 100°F are classified as flammable and should be handled with special precautions to prevent combustion.

7. First Aid – Gives a letter reference as to the measures that should be taken when an accidental contamination of the eyes, skin, lungs, or intestines has occurred. See Tables 2-3-1 through 2-3-4 for an index of the referenced letters.

8. Spill – Gives a letter reference as to the measures to be taken when an accidental spill occurs. See Table 2-4 for an index of the referenced letters.

9. Disposal – Gives a letter reference as to the measures to be taken for the proper disposal of the material listed in the second column. See Table 2-5 for an index of the referenced letters.

10. Location – Indicates the location where the product may be found within the facility.

D. Index for Workplace Chemical List - The following tables explain and/or describe the letters assigned to the different rows of the Workplace Chemical List. Included are indices for personal protective equipment, first aid, spill cleanup, and disposal measures. These tables may be found in the front of the Workplace Chemical List at each listed location. There is also a Personal Protection Index posted in various locations throughout all facilities where hazardous materials are present.

## Table 2-1: Hazard Index

**Health Hazard (Blue)**

4 - Deadly; short exposure may cause death or serious injury

3 - Extremely dangerous; short exposure could cause serious, temporary, or residual injury

2 - Hazardous; intense or continued exposure could cause temporary incapacitation or possible residual injury

1 - Slightly hazardous; on exposure, could cause irritation, but only minor residual injury

0 - No hazard

**Reactivity Hazard (Yellow)**

4 - Extremely reactive, may detonate

3 - May detonate with shock and heat

2 - Chemical change may be violent

1 - Unstable if heated

0 - Stable

**Fire Hazard (Red)**

4 - Extremely flammable, flash point below 0°F

3 - Highly flammable, flash point 1°F-100°F

2 - Moderately flammable, flash point 101°F-200°F

1 - Slightly flammable, flash point above 201°F

0 - Not flammable

**Basic Hazard Index**

4 - Severe Hazard

3 - Serious Hazard

2 - Moderate hazard

1 - Slight Hazard

0 - Minimal Hazard

## Table 2-2: Personal Protection Index

| | |
|---|---|
| A. | Safety Glasses |
| B. | Safety Glasses, Chemical Gloves |
| C. | Safety Glasses, Chemical Gloves, Synthetic Apron |
| D. | Face Shield, Chemical Gloves, Synthetic Apron |
| E. | Safety Glasses, Chemical Gloves, Dust Respirator |
| F. | Safety Glasses, Chemical Gloves, Synthetic Apron, Dust Respirator |
| G. | Safety Glasses, Chemical Gloves, Vapor Respirator |
| H. | Splash Goggles, Chemical Gloves, Synthetic Apron, Vapor Respirator |
| I. | Safety Glasses, Chemical Gloves, Dust and Vapor Respirator |
| J. | Splash Goggles, Chemical Gloves, Synthetic Apron, Dust and Vapor Respirator |
| K. | Airline Hood or Mask, Chemical Gloves, Full Suit, Rubber Boots |
| X. | Ask your supervisor for special handling instructions |

First Aid – Tables 2-3-1 through 2-3-4 define the emergency procedures recommended for the treatment of an individual who has experienced an injury or exposure to a particular hazardous material. These first aid measures have been taken from the MSDSs and are considered as interim procedures only to be administered between exposure and treatment by licensed/trained medical personnel. These measures in no way are intended to replace or supersede professional treatment in the case of injury or exposure. A licensed physician should examine an injured employee as soon as possible.

## Table 2-3-1: First Aid – Eyes

| | |
|---|---|
| A. | Flush with water for 15 minutes. Get medical attention. |
| B. | Flush with plenty of water. |
| C. | Flush with lukewarm water for 15 minutes. Get medical attention. |
| D. | Flush with water until irritation subsides. |
| E. | Flush with warm water. Apply gauze patch. Get medical attention. |
| F. | Flush with cold water. Get medical attention. |
| G. | Flush with water for 15 minutes. |
| H. | No first aid should be needed. |
| X. | Reference respective MSDS number for specific first aid information. |

## Table 2-3-2: First Aid – Skin

| | |
|---|---|
| A. | Wash with soap and water. Contact physician if irritation persists. |
| B. | Wash with soap and water. Apply skin cream that contains lanolin. |
| C. | Flush with water for 15 minutes. |
| D. | Flush with water. Contact physician if irritation persists. |
| E. | Soak in warm soapy water. |
| F. | Wipe off product and flush with water. |
| G. | Wash with soap and water. Get medical attention. |
| H. | Wipe off contamination. Wash with soap and water. |
| I. | Flush with lukewarm water. |
| J. | No first aid should be needed. |
| X. | Reference respective MSDS number for specific first aid information. |

## Table 2-3-3: First Aid – Inhalation

A. Remove to fresh air. If breathing has stopped, give mouth-to-mouth. Get medical attention.
B. Remove to fresh air. Keep warm and quiet. Get medical attention.
C. No first aid treatment should be necessary.
X. Reference respective MSDS number for specific first-aid information.

## Table 2-3-4: First Aid – Ingestion

A. Do not induce vomiting. Keep warm and quiet. Get medical attention.
B. Rinse mouth. Give plenty of water followed by milk, egg white, or gruel. Do not induce vomiting. Get medical attention.
C. Contact poison control center or physician immediately.
D. Give plenty of water. Get medical attention.
E. Give two (2) glasses of water. Induce vomiting. Get medical attention.
F. Do not induce vomiting. Give plenty of water. Get medical attention.
G. Give milk, water, or egg whites. Induce vomiting. Get medical attention.
H. Induce vomiting by giving a tablespoon of salt or mustard in a glass of warm water. Get medical attention.
I. Induce vomiting. Get medical attention.
J. Drink Milk of Magnesia, aluminum hydroxide gel, or lime water, followed by several glasses of water. Call physician. Do not induce vomiting.
K. Do not induce vomiting. Drink two (2) glasses of milk or water. Call physician or poison control center.
L. Rinse mouth with water several times.
M. No first aid should be needed.
X. Reference respective MSDS number for specific first aid information.

Spills – Table 2-4 describes the measures that should be taken to properly contain and clean up any material spilled.

## Table 2-4: Spill Containment and Cleanup

A. Absorb on inert material.
B. Eliminate ignition sources. Ventilate area. Absorb on inert material.
C. Absorb on inert material. Flush area with water.
D. Dike spill. Absorb on inert material. Sprinkle sodium thiosulfate on residue. Remove residue. Clean area with water.
E. Vacuum, scoop, or sweep discharged material.
F. Dilute by adding large volumes of water.
G. Neutralize with lime or soda ash. Absorb on inert material.
H. No spill or clean-up measures should be necessary.
X. Reference respective MSDS number for specific spill containment and clean-up procedures.

Disposal – Table 2-5 defines procedures for waste disposal of the materials listed.

## Table 2-5: Waste Disposal Procedures

| | |
|---|---|
| A. | Consult local, state, and federal regulations. |
| B. | Do not incinerate container. Consult local, state, and federal regulations. |
| C. | Flush to sewer. |
| D. | There should be no residue for disposal. |
| E. | Dispose in trash. |

*Note:* *The EPA levies extremely high civil penalties for the improper disposal of hazardous/toxic substances. Consult local waste treatment control and disposal center and ensure compliance with all local, state, and federal regulations.*

**10. HAZARDOUS MATERIALS LABELING**: Federal regulations require all containers used in the storage and/or transportation of hazardous and toxic chemicals to be labeled, tagged, or marked with the following information:

A. Identity of the hazardous chemical(s);

B. Appropriate hazard warning; and,

C. The name and address of the manufacturer (for products received only).

In some cases, this regulation is satisfied by the warning information that is printed on the product label. However, there are some products that do not have this information. In such cases, the department manager/supervisor will be responsible for identifying and labeling such products.

The department manager/supervisor will be responsible for ensuring that all portable containers are labeled. Portable containers are defined as pressure sprayers, spray bottles, squeeze bottles, hydraulic servicing units, or any container used in the transfer or temporary storage of products which contain hazardous ingredients as defined in Title 29 CFR 1910.1200.

The Hazardous Materials Identification System (HMIS) is a standard system that communicates the hazards to the employee. The Hazardous Materials Identification Guide (HMIG), as it is referred to, communicates necessary hazard information by use of a common numerical system for Health (blue), Flammability (red), Reactivity (yellow), and Personal Protective Equipment required or specific hazard information (white).

In order to properly complete the HMIG label (see Figure 2-3), it is necessary to reference the Workplace Chemical List. In the list, find the product that is to be labeled. Indicate the name of the product on the top section of the label (if not legible on container). The Health, Flammability, and Reactivity Section is rated by using the numerical ratings indicated in the Hazard Rating row. To indicate the recommended Personal Protective Equipment, use the letter that is given in the PPE row.

## Figure 2-3: Hazardous Materials Identification Guide

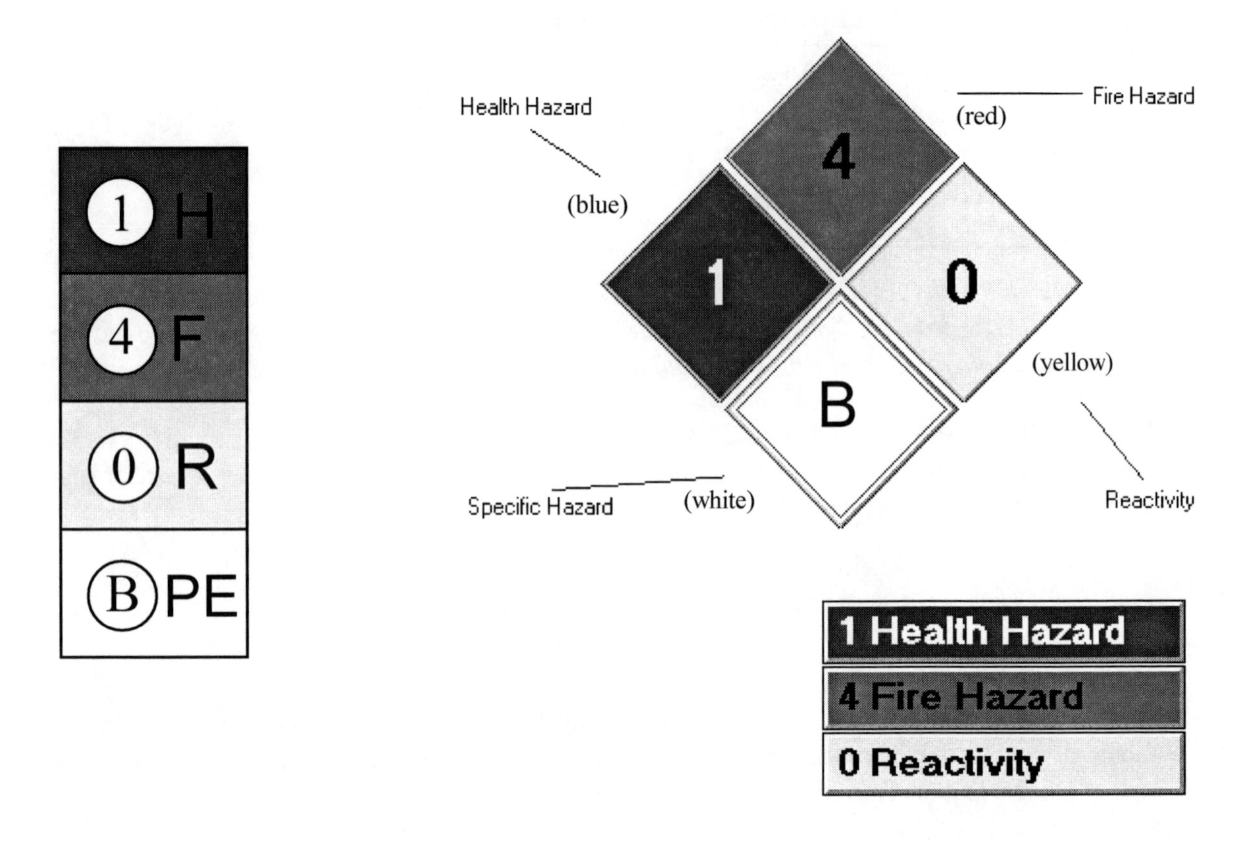


For an explanation of the Hazard Index, refer to Table 2-1. The explanation for the Personal Protection Index may be found in Table 2-2.

11. **HEALTH HAZARD CHARACTERIZATION**: Materials that have one or more of the following characteristic properties are covered by the Hazard Communication Standard:

| | | |
|---|---|---|
| irritations | cutaneous hazards | toxic agents |
| highly toxic agents | corrosives | eye hazards |
| blood acting agents | sensitizers | carcinogens |
| hematopietic system agents | hepatoxins | neurotoxins |
| reproductive toxins | nephrotoxins | |

agents that damage the lungs, skin, or mucous membranes

Health hazards are divided between acute health hazards and chronic health hazards. Some hazards have characteristics of both, but generally they are one or the other.

A. Acute Hazards - The effects of acute hazards are manifested soon after a single, brief exposure. Many acute effects disappear after a time and generally are not permanent. However, some may show permanent effects, and therefore can be considered both acute and chronic.

Acute, when used to describe a hazard, does not imply severity. For the purpose of this program, the term acute refers to a short-term effect.

1. Irritants - An irritant is defined as a chemical capable of causing reversible inflammation at the site of contact by chemical action. Examples include nitric oxide, sodium hypochlorite, stannic chloride, and ethyl alcohol.

2. Cutaneous Hazards - A cutaneous hazard is a material that affects the dermal layer of the body, such as defatting of the skin and causing rashes or skin irritation. Examples include acetone and chlorinated compounds.

3. Toxic Agents - Toxic agents are those defined as substances which can cause acute injury to the human body, or which are suspected of being able to cause diseases or injury under certain conditions.

4. Corrosive Materials - A corrosive material is a chemical causing visible destruction of, or irreversible alterations in living tissue at the site of contact by chemical reaction. Examples include caustic soda, sulfuric acid, hydrofluoric acid, phenol, and boron trifluoride.

5. Eye Hazards - An eye hazard is a material that affects the eye or visual capacity, for example, by causing conjunctivitis or corneal damage. Examples include organic solvents, acids, and alkalis.

6. Agents that Act on the Blood or Hemalopoietic System - This type of agent is a substance that decreases the hemoglobin function and deprives the body tissue of oxygen. Cyanosis and loss of consciousness are typical symptoms. Examples include carbon monoxide, cyanides, metal carbonyls, nitrobenzene, hydroquinone, aniline, and arsine.

B. Chronic Hazards - Chronic hazards have a long-term affect, essentially permanent. Their affects may be slow to develop, and often result from repeated or continuous exposure over a long period of time.

1. Sensitizers - A sensitizer is a chemical that causes a substantial portion of exposed people or animals to develop an allergic reaction in normal tissue after repeated exposure. Examples include hydroquinone, bromine, platinum compounds, isocyanates, and ozone.

2. Carcinogens - A substance or agent capable of causing or producing cancer in mammals, including humans. Examples include asbestos, benzene, beryllium, lead chromate, formaldehyde, vinyl chloride, trichloroethylene, and carbon tetrachloride.

3. Reproductive Toxins (teratogens) - Substances that can cause birth defects or sterility.

4. Hepatoxins - A hepatoxin is a chemical that can cause liver damage such as enlargement or jaundice. Examples include carbon tetrachloride, nitrosamines, vinyl chloride, chlorobenzene, trichloroethylene, chloroform, and ethyl alcohol.

5. Nephrotoxins - A nephrotoxin is a chemical that can cause kidney damage such as edema or proteinuria. Examples include halogenated hydrocarbons, uranium, vinyl chloride, trichloroethylene, and ethyl alcohol.

6. Neurotoxins - A neurotoxin is a chemical that causes primary toxic affects on the central nervous system, such as narcosis, behavioral changes, or a decrease in motor functions. Examples include mercury, carbon disulfide, ethyl alcohol, acetylene, manganese, thallium, and tetraethyl lead.

7. Agents that Damage the Lungs - These agents irritate pulmonary tissue, resulting in cough, tightness in the chest, and shortness of breath. Examples include silica, asbestos, cotton fibers, coal dust, and toluene diisocyanate.

12. **"RIGHT-TO-KNOW" POSTER**: A "Right-to-Know" poster must be displayed at each location where products are used that are considered hazardous by OSHA and state regulations. The following are good places for posters to be displayed:

A. Bulletin boards in main operations and employee break areas

B. Shop bulletin boards

C. Maintenance areas

**NOTE:** The "Right-to-Know" poster may be ordered from _______________ or from a local safety supply company. Please indicate the number required.

# 1910.1200 Hazard Communication Standard

*The following pages contain the Hazard Communication Standard exactly as it can be found in Title 29 of the Code of Federal Regulations, Part 1910.1200. You will quickly notice that the information in this standard can be difficult to follow. However, it should be read before implementation of your HazCom program to assure no additional requirements may be applicable to your facility and to assure a complete understanding of the standard as it relates to your facility.*

(a) "Purpose."

(1) The purpose of this section is to ensure that the hazards of all chemicals produced or imported are evaluated, and that information concerning their hazards is transmitted to employers and employees. This transmittal of information is to be accomplished by means of comprehensive hazard communication programs, which are to include container labeling and other forms of warning, material safety data sheets and employee training.

(2) This occupational safety and health standard is intended to address comprehensively the issue of evaluating the potential hazards of chemicals, and communicating information concerning hazards and appropriate protective measures to employees, and to preempt any legal requirements of a state, or political subdivision of a state, pertaining to this subject. Evaluating the potential hazards of chemicals, and communicating information concerning hazards and appropriate protective measures to employees, may include, for example, but is not limited to, provisions for: developing and maintaining a written hazard communication program for the workplace, including lists of hazardous chemicals present; labeling of containers of chemicals in the workplace, as well as of containers of chemicals being shipped to other workplaces; preparation and distribution of material safety data sheets to employees and downstream employers; and development and implementation of employee training programs regarding hazards of chemicals and protective measures. Under section 18 of the Act, no state or political subdivision of a state may adopt or enforce, through any court or agency, any requirement relating to the issue addressed by this Federal standard, except pursuant to a Federally-approved state plan.

(b) "Scope and application."

(1) This section requires chemical manufacturers or importers to assess the hazards of chemicals which they produce or import, and all employers to provide information to their employees about the hazardous chemicals to which they are exposed, by means of a hazard communication program, labels and other forms of warning, material safety data sheets, and information and training. In addition, this section requires distributors to transmit the required information to employers. (Employers who do not produce or import chemicals need only focus on those parts of this rule that deal with establishing a workplace program and communicating information to their workers. Appendix E of this section is a general guide for such employers to help them determine their compliance obligations under the rule.)

(2) This section applies to any chemical which is known to be present in the workplace in such a manner that employees may be exposed under normal conditions of use or in a foreseeable emergency.

(3) This section applies to laboratories only as follows:

(i) Employers shall ensure that labels on incoming containers of hazardous chemicals are not removed or defaced;

(ii) Employers shall maintain any material safety data sheets that are received with incoming shipments of hazardous chemicals, and ensure that they are readily accessible during each workshift to laboratory employees when they are in their work areas;

(iii) Employers shall ensure that laboratory employees are provided information and training in accordance with paragraph (h) of this section, except for the location and availability of the written hazard communication program under paragraph (h)(2)(iii) of this section; and,

(iv) Laboratory employers that ship hazardous chemicals are considered to be either a chemical manufacturer or a distributor under this rule, and thus must ensure that any containers of hazardous chemicals leaving the laboratory are labeled in accordance with paragraph (f)(1) of this section, and that a material safety data sheet is provided to distributors and other employers in accordance with paragraphs (g)(6) and (g)(7) of this section.

(4) In work operations where employees only handle chemicals in sealed containers which are not opened under normal conditions of use (such as are found in marine cargo handling, warehousing, or retail sales), this section applies to these operations only as follows:

(i) Employers shall ensure that labels on incoming containers of hazardous chemicals are not removed or defaced;

(ii) Employers shall maintain copies of any material safety data sheets that are received with incoming shipments of the sealed containers of hazardous chemicals, shall obtain a material safety data sheet as soon as possible for sealed containers of hazardous chemicals received without a material safety data sheet if an employee requests the material safety data sheet, and shall ensure that the material safety data sheets are readily accessible during each work shift to employees when they are in their work area(s); and,

(iii) Employers shall ensure that employees are provided with information and training in accordance with paragraph (h) of this section (except for the location and availability of the written hazard communication program under paragraph (h)(2)(iii) of this section), to the extent necessary to protect them in the event of a spill or leak of a hazardous chemical from a sealed container.

(5) This section does not require labeling of the following chemicals:

(i) Any pesticide as such term is defined in the Federal Insecticide, Fungicide, and Rodenticide Act (7 U.S.C. 136 et seq.), when subject to the labeling requirements of that Act and labeling regulations issued under that Act by the Environmental Protection Agency;

(ii) Any chemical substance or mixture as such terms are defined in the Toxic Substances Control Act (15 U.S.C. 2601 et seq.), when subject to the labeling requirements of that Act and labeling regulations issued under that Act by the Environmental Protection Agency;

(iii) Any food, food additive, color additive, drug, cosmetic, or medical or veterinary device or product, including materials intended for use as ingredients in such products (e.g. flavors and fragrances), as such terms are defined in the Federal Food, Drug, and Cosmetic Act (21 U.S.C. 301 et seq.) or the Virus-Serum-Toxin Act of 1913 (21 U.S.C. 151 et seq.), and regulations issued under those Acts, when they are subject to the labeling requirements under those Acts by either the Food and Drug Administration or the Department of Agriculture;

(iv) Any distilled spirits (beverage alcohols), wine, or malt beverage intended for nonindustrial use, as such terms are defined in the Federal Alcohol Administration Act (27 U.S.C. 201 et seq.) and regulations issued under that Act, when subject to the labeling requirements of that Act and labeling regulations issued under that Act by the Bureau of Alcohol, Tobacco, and Firearms;

(v) Any consumer product or hazardous substance as those terms are defined in the Consumer Product Safety Act (15 U.S.C. 2051 et seq.) and Federal Hazardous Substances Act (15 U.S.C. 1261 et seq.) respectively, when subject to a consumer product safety standard or labeling requirement of those Acts, or regulations issued under those Acts by the Consumer Product Safety Commission; and,

(vi) Agricultural or vegetable seed treated with pesticides and labeled in accordance with the Federal Seed Act (7 U.S.C. 1551 et seq.) and the labeling regulations issued under that Act by the Department of Agriculture.

(6) This section does not apply to:

(i) Any hazardous waste as such term is defined by the Solid Waste Disposal Act, as amended by the Resource Conservation and Recovery Act of 1976, as amended (42 U.S.C. 6901 et seq.), when subject to regulations issued under that Act by the Environmental Protection Agency;

(ii) Any hazardous substance as such term is defined by the Comprehensive Environmental Response, Compensation and Liability ACT (CERCLA) (42 U.S.C. 9601 et seq.) when the hazardous substance is the focus of remedial or removal action being conducted under CERCLA in accordance with the Environmental Protection Agency regulations.

(iii) Tobacco or tobacco products;

(iv) Wood or wood products, including lumber which will not be processed, where the chemical manufacturer or importer can establish that the only hazard they pose to employees is the potential for flammability or combustibility (wood or wood products which have been treated with a hazardous chemical covered by this standard, and wood which may be subsequently sawed or cut, generating dust, are not exempted);

(v) Articles (as that term is defined in paragraph (c) of this section);

(vi) Food or alcoholic beverages which are sold, used, or prepared in a retail establishment (such as a grocery store, restaurant, or drinking place), and foods intended for personal consumption by employees while in the workplace;

(vii) Any drug, as that term is defined in the Federal Food, Drug, and Cosmetic Act (21 U.S.C. 301 et seq.), when it is in solid, final form for direct administration to the patient (e.g., tablets or pills); drugs which are packaged by the chemical manufacturer for sale to consumers in a retail establishment (e.g., over-the-counter drugs); and drugs intended for personal consumption by employees while in the workplace (e.g., first aid supplies);

(viii) Cosmetics which are packaged for sale to consumers in a retail establishment, and cosmetics intended for personal consumption by employees while in the workplace;

(ix) Any consumer product or hazardous substance, as those terms are defined in the Consumer Product Safety Act (15 U.S.C. 2051 et seq.) and Federal Hazardous Substances Act (15 U.S.C. 1261 et seq.) respectively, where the employer can show that it is used in the workplace for the purpose intended by the chemical manufacturer or importer of the product, and the use results in a duration and frequency of exposure which is not greater than the range of exposures that could reasonably be experienced by consumers when used for the purpose intended;

(x) Nuisance particulates where the chemical manufacturer or importer can establish that they do not pose any physical or health hazard covered under this section;

(xi) Ionizing and nonionizing radiation; and,

(xii) Biological hazards.

(c) "Definitions."

"**Article**" means a manufactured item other than a fluid or particle:

(i) which is formed to a specific shape or design during manufacture;

(ii) which has end use function(s) dependent in whole or in part upon its shape or design during end use; and

(iii) which under normal conditions of use does not release more than very small quantities, e.g., minute or trace amounts of a hazardous chemical (as determined under paragraph (d) of this section), and does not pose a physical hazard or health risk to employees.

"**Assistant Secretary**" means the Assistant Secretary of Labor for Occupational Safety and Health, U.S. Department of Labor, or designee.

"**Chemical**" means any element, chemical compound or mixture of elements and/or compounds.

"**Chemical manufacturer**" means an employer with a workplace where chemical(s) are produced for use or distribution.

"**Chemical name**" means the scientific designation of a chemical in accordance with the nomenclature system developed by the International Union of Pure and Applied Chemistry (IUPAC) or the Chemical Abstracts Service (CAS) rules of nomenclature, or a name which will clearly identify the chemical for the purpose of conducting a hazard evaluation.

"**Combustible liquid**" means any liquid having a flashpoint at or above 100 deg. F (37.8 deg. C), but below 200 deg. F (93.3 deg. C), except any mixture having components with flashpoints of 200 deg. F (93.3 deg. C), or higher, the total volume of which make up 99 percent or more of the total volume of the mixture.

"**Commercial account**" means an arrangement whereby a retail distributor sells hazardous chemicals to an employer, generally in large quantities over time and/or at costs that are below the regular retail price.

"**Common name**" means any designation or identification such as code name, code number, trade name, brand name or generic name used to identify a chemical other than by its chemical name.

"**Compressed gas**" means:

(i) A gas or mixture of gases having, in a container, an absolute pressure exceeding 40 psi at 70 deg. F (21.1 deg. C); or

(ii) A gas or mixture of gases having, in a container, an absolute pressure exceeding 104 psi at 130 deg. F (54.4 deg. C) regardless of the pressure at 70 deg. F (21.1 deg. C); or

(iii) A liquid having a vapor pressure exceeding 40 psi at 100 deg. F (37.8 deg. C) as determined by ASTM D-323-72.

"**Container**" means any bag, barrel, bottle, box, can, cylinder, drum, reaction vessel, storage tank, or the like that contains a hazardous chemical. For purposes of this section, pipes or piping systems, and engines, fuel tanks, or other operating systems in a vehicle, are not considered to be containers.

"**Designated representative**" means any individual or organization to whom an employee gives written authorization to exercise such employee's rights under this section. A recognized or certified collective bargaining agent shall be treated automatically as a designated representative without regard to written employee authorization.

"**Director**" means the Director, National Institute for Occupational Safety and Health, U.S. Department of Health and Human Services, or designee.

"**Distributor**" means a business, other than a chemical manufacturer or importer, which supplies hazardous chemicals to other distributors or to employers.

"**Employee**" means a worker who may be exposed to hazardous chemicals under normal operating conditions or in foreseeable emergencies. Workers such as office workers or bank tellers who encounter hazardous chemicals only in non-routine, isolated instances are not covered.

"**Employer**" means a person engaged in a business where chemicals are either used, distributed, or are produced for use or distribution, including a contractor or subcontractor.

"**Explosive**" means a chemical that causes a sudden, almost instantaneous release of pressure, gas, and heat when subjected to sudden shock, pressure, or high temperature.

"**Exposure or exposed**" means that an employee is subjected in the course of employment to a chemical that is a physical or health hazard, and includes potential (e.g. accidental or possible) exposure. "Subjected" in terms of health hazards includes any route of entry (e.g. inhalation, ingestion, skin contact or absorption.)

"**Flammable**" means a chemical that falls into one of the following categories:

(i) "**Aerosol, flammable**" means an aerosol that, when tested by the method described in 16 CFR 1500.45, yields a flame projection exceeding 18 inches at full valve opening, or a flashback (a flame extending back to the valve) at any degree of valve opening;

(ii) "**Gas, flammable**" means:

(A) A gas that, at ambient temperature and pressure, forms a flammable mixture with air at a concentration of thirteen (13) percent by volume or less; or

(B) A gas that, at ambient temperature and pressure, forms a range of flammable mixtures with air wider than twelve (12) percent by volume, regardless of the lower limit;

(iii) "**Liquid, flammable**" means any liquid having a flashpoint below 100 deg. F (37.8 deg. C), except any mixture having components with flashpoints of 100 deg. F (37.8 deg. C) or higher, the total of which make up 99 percent or more of the total volume of the mixture.

(iv) "**Solid, flammable**" means a solid, other than a blasting agent or explosive as defined in 1910.109(a), that is liable to cause fire through friction, absorption of moisture, spontaneous chemical change, or retained heat from manufacturing or processing, or which can be ignited readily and when ignited burns so vigorously and persistently as to create a serious hazard. A chemical shall be considered to be a flammable solid if, when tested by the method described in 16 CFR 1500.44, it ignites and burns with a self-sustained flame at a rate greater than one-tenth of an inch per second along its major axis.

"**Flashpoint**" means the minimum temperature at which a liquid gives off a vapor in sufficient concentration to ignite when tested as follows:

(i) Tagliabue Closed Tester (See American National Standard Method of Test for Flash Point by Tag Closed Tester, Z11.24-1979 (ASTM D 56-79)) for liquids with a viscosity of less than 45 Saybolt Universal Seconds (SUS) at 100 deg. F (37.8 deg. C), that do not contain suspended solids and do not have a tendency to form a surface film under test; or

(ii) Pensky-Martens Closed Tester (see American National Standard Method of Test for Flash Point by Pensky-Martens Closed Tester, Z11.7-1979 (ASTM D 93-79)) for liquids with a viscosity equal to or greater than 45 SUS at 100 deg. F (37.8 deg. C), or that contain suspended solids, or that have a tendency to form a surface film under test; or

(iii) Setaflash Closed Tester (see American National Standard Method of Test for Flash Point by Setaflash Closed Tester (ASTM D 3278-78)).

Organic peroxides, which undergo autoaccelerating thermal decomposition, are excluded from any of the flashpoint determination methods specified above.

"**Foreseeable emergency**" means any potential occurrence such as, but not limited to, equipment failure, rupture of containers, or failure of control equipment which could result in an uncontrolled release of a hazardous chemical into the workplace.

"**Hazardous chemical**" means any chemical which is a physical hazard or a health hazard.

"**Hazard warning**" means any words, pictures, symbols, or combination thereof appearing on a label or other appropriate form of warning which convey the specific physical and health hazard(s), including target organ effects, of the chemical(s) in the container(s). (See the definitions for "physical hazard" and "health hazard" to determine the hazards which must be covered.)

"**Health hazard**" means a chemical for which there is statistically significant evidence based on at least one study conducted in accordance with established scientific principles that acute or chronic health effects may occur in exposed employees. The term "health hazard" includes chemicals which are carcinogens, toxic or highly toxic agents, reproductive toxins, irritants, corrosives, sensitizers, hepatotoxins, nephrotoxins, neurotoxins, agents which act on the hematopoietic system, and agents which damage the lungs, skin, eyes, or mucous membranes. Appendix A provides further definitions and explanations of the scope of health hazards covered by this section, and Appendix B describes the criteria to be used to determine whether or not a chemical is to be considered hazardous for purposes of this standard.

"**Identity**" means any chemical or common name which is indicated on the material safety data sheet (MSDS) for the chemical. The identity used shall permit cross-references to be made among the required list of hazardous chemicals, the label and the MSDS.

"**Immediate use**" means that the hazardous chemical will be under the control of and used only by the person who transfers it from a labeled container and only within the work shift in which it is transferred.

"**Importer**" means the first business with employees within the Customs Territory of the United States which receives hazardous chemicals produced in other countries for the purpose of supplying them to distributors or employers within the United States.

"**Label**" means any written, printed, or graphic material displayed on or affixed to containers of hazardous chemicals.

"**Material safety data sheet (MSDS)**" means written or printed material concerning a hazardous chemical which is prepared in accordance with paragraph (g) of this section.

"**Mixture**" means any combination of two or more chemicals if the combination is not, in whole or in part, the result of a chemical reaction.

"**Organic peroxide**" means an organic compound that contains the bivalent -O-O-structure and which may be considered to be a structural derivative of hydrogen peroxide where one or both of the hydrogen atoms has been replaced by an organic radical.

"**Oxidizer**" means a chemical other than a blasting agent or explosive as defined in 1910.109(a), that initiates or promotes combustion in other materials, thereby causing fire either of itself or through the release of oxygen or other gases.

"**Physical hazard**" means a chemical for which there is scientifically valid evidence that it is a combustible liquid, a compressed gas, explosive, flammable, an organic peroxide, an oxidizer, pyrophoric, unstable (reactive) or water-reactive.

"**Produce**" means to manufacture, process, formulate, blend, extract, generate, emit, or repackage.

"**Pyrophoric**" means a chemical that will ignite spontaneously in air at a temperature of 130 deg. F (54.4 deg. C) or below.

"**Responsible party**" means someone who can provide additional information on the hazardous chemical and appropriate emergency procedures, if necessary.

"**Specific chemical identity**" means the chemical name, Chemical Abstracts Service (CAS) Registry Number, or any other information that reveals the precise chemical designation of the substance.

"**Trade secret**" means any confidential formula, pattern, process, device, information or compilation of information that is used in an employer's business, and that gives the employer an opportunity to obtain an advantage over competitors who do not know or use it. Appendix D sets out the criteria to be used in evaluating trade secrets.

**"Unstable (reactive)"** means a chemical which in the pure state, or as produced or transported, will vigorously polymerize, decompose, condense, or will become self-reactive under conditions of shocks, pressure or temperature.

**"Use"** means to package, handle, react, emit, extract, generate as a byproduct, or transfer.

**"Water-reactive"** means a chemical that reacts with water to release a gas that is either flammable or presents a health hazard.

**"Work area"** means a room or defined space in a workplace where hazardous chemicals are produced or used, and where employees are present.

**"Workplace"** means an establishment, job site, or project, at one geographical location containing one or more work areas.

(d) "Hazard determination."

(1) Chemical manufacturers and importers shall evaluate chemicals produced in their workplaces or imported by them to determine if they are hazardous. Employers are not required to evaluate chemicals unless they choose not to rely on the evaluation performed by the chemical manufacturer or importer for the chemical to satisfy this requirement.

(2) Chemical manufacturers, importers or employers evaluating chemicals shall identify and consider the available scientific evidence concerning such hazards. For health hazards, evidence which is statistically significant and which is based on at least one positive study conducted in accordance with established scientific principles is considered to be sufficient to establish a hazardous effect if the results of the study meet the definitions of health hazards in this section. Appendix A shall be consulted for the scope of health hazards covered, and Appendix B shall be consulted for the criteria to be followed with respect to the completeness of the evaluation, and the data to be reported.

(3) The chemical manufacturer, importer or employer evaluating chemicals shall treat the following sources as establishing that the chemicals listed in them are hazardous:

(i) 29 CFR part 1910, subpart Z, Toxic and Hazardous Substances, Occupational Safety and Health Administration (OSHA); or,

(ii) "Threshold Limit Values for Chemical Substances and Physical Agents in the Work Environment," American Conference of Governmental Industrial Hygienists (ACGIH) (latest edition). The chemical manufacturer, importer, or employer is still responsible for evaluating the hazards associated with the chemicals in these source lists in accordance with the requirements of this standard.

(4) Chemical manufacturers, importers and employers evaluating chemicals shall treat the following sources as establishing that a chemical is a carcinogen or potential carcinogen for hazard communication purposes:

(i) National Toxicology Program (NTP), "Annual Report on Carcinogens" (latest edition);

(ii) International Agency for Research on Cancer (IARC) "Monographs" (latest editions); or

(iii) 29 CFR part 1910, subpart Z, Toxic and Hazardous Substances, Occupational Safety and Health Administration.

Note: The "Registry of Toxic Effects of Chemical Substances" published by the National Institute for Occupational Safety and Health indicates whether a chemical has been found by NTP or IARC to be a potential carcinogen.

(5) The chemical manufacturer, importer or employer shall determine the hazards of mixtures of chemicals as follows:

(i) If a mixture has been tested as a whole to determine its hazards, the results of such testing shall be used to determine whether the mixture is hazardous;

(ii) If a mixture has not been tested as a whole to determine whether the mixture is a health hazard, the mixture shall be assumed to present the same health hazards as do the components which comprise one percent (by weight or volume) or greater of the mixture, except that the mixture shall be assumed to present a carcinogenic hazard if it contains a component in concentrations of 0.1 percent or greater which is considered to be a carcinogen under paragraph (d)(4) of this section;

(iii) If a mixture has not been tested as a whole to determine whether the mixture is a physical hazard, the chemical manufacturer, importer, or employer may use whatever scientifically valid data is available to evaluate the physical hazard potential of the mixture; and,

(iv) If the chemical manufacturer, importer, or employer has evidence to indicate that a component present in the mixture in concentrations of less than one percent (or in the case of carcinogens, less than 0.1 percent) could be released in concentrations which would exceed an established OSHA permissible exposure limit or ACGIH Threshold Limit Value, or could present a health risk to employees in those concentrations, the mixture shall be assumed to present the same hazard.

(6) Chemical manufacturers, importers, or employers evaluating chemicals shall describe in writing the procedures they use to determine the hazards of the chemical they evaluate. The written procedures are to be made available, upon request, to employees, their designated representatives, the Assistant Secretary and the Director. The written description may be incorporated into the written hazard communication program required under paragraph (e) of this section.

1910.1200(e)(1) Written Hazard Communication Program - Training Required

(e) "Written hazard communication program."

(1) Employers shall develop, implement, and maintain at each workplace, a written hazard communication program which at least describes how the criteria specified in paragraphs (f), (g), and (h) of this section for labels and other forms of warning, material safety data sheets, and employee information and training will be met, and which also includes the following:

(i) A list of the hazardous chemicals known to be present using an identity that is referenced on the appropriate material safety data sheet (the list may be compiled for the workplace as a whole or for individual work areas); and,

(ii) The methods the employer will use to inform employees of the hazards of non-routine tasks (for example, the cleaning of reactor vessels), and the hazards associated with chemicals contained in unlabeled pipes in their work areas.

(2) "Multi-employer workplaces." Employers who produce, use, or store hazardous chemicals at a workplace in such a way that the employees of other employer(s) may be exposed (for example, employees of a construction contractor working on-site) shall additionally ensure that the hazard communication programs developed and implemented under this paragraph (e) include the following:

(i) The methods the employer will use to provide the other employer(s) on-site access to material safety data sheets for each hazardous chemical the other employer(s)' employees may be exposed to while working;

(ii) The methods the employer will use to inform the other employer(s) of any precautionary measures that need to be taken to protect employees during the workplace's normal operating conditions and in foreseeable emergencies; and,

(iii) The methods the employer will use to inform the other employer(s) of the labeling system used in the workplace.

(3) The employer may rely on an existing hazard communication program to comply with these requirements, provided that it meets the criteria established in this paragraph (e).

(4) The employer shall make the written hazard communication program available, upon request, to employees, their designated representatives, the Assistant Secretary and the Director, in accordance with the requirements of 29 CFR 1910.1020 (e).

(5) Where employees must travel between workplaces during a workshift, i.e., their work is carried out at more than one geographical location, the written hazard communication program may be kept at the primary workplace facility.

(f) "Labels and other forms of warning."

(1) The chemical manufacturer, importer, or distributor shall ensure that each container of hazardous chemicals leaving the workplace is labeled, tagged or marked with the following information:

(i) Identity of the hazardous chemical(s);

(ii) Appropriate hazard warnings; and,

(iii) Name and address of the chemical manufacturer, importer, or other responsible party.

(2)(i) For solid metal (such as a steel beam or a metal casting), solid wood, or plastic items that are not exempted as articles due to their downstream use, or shipments of whole grain, the required label may be transmitted to the customer at the time of the initial shipment, and need not be included with subsequent shipments to the same employer unless the information on the label changes;

(ii) The label may be transmitted with the initial shipment itself, or with the material safety data sheet that is to be provided prior to or at the time of the first shipment; and,

(iii) This exception to requiring labels on every container of hazardous chemicals is only for the solid material itself, and does not apply to hazardous chemicals used in conjunction with, or known to be present with, the material and to which employees handling the items in transit may be exposed (for example, cutting fluids or pesticides in grains).

(3) Chemical manufacturers, importers, or distributors shall ensure that each container of hazardous chemicals leaving the workplace is labeled, tagged, or marked in accordance with this section in a manner which does not conflict with the requirements of the Hazardous Materials Transportation Act (49 U.S.C. 1801 et seq.) and regulations issued under that Act by the Department of Transportation.

(4) If the hazardous chemical is regulated by OSHA in a substance-specific health standard, the chemical manufacturer, importer, distributor or employer shall ensure that the labels or other forms of warning used are in accordance with the requirements of that standard.

(5) Except as provided in paragraphs (f)(6) and (f)(7) of this section, the employer shall ensure that each container of hazardous chemicals in the workplace is labeled, tagged or marked with the following information:

(i) Identity of the hazardous chemical(s) contained therein; and,

(ii) Appropriate hazard warnings, or alternatively, words, pictures, symbols, or combination thereof, which provide at least general information regarding the hazards of the chemicals, and which, in conjunction with the other information immediately available to employees under the hazard communication program, will provide employees with the specific information regarding the physical and health hazards of the hazardous chemical.

(6) The employer may use signs, placards, process sheets, batch tickets, operating procedures, or other such written materials in lieu of affixing labels to individual stationary process containers, as long as the alternative method identifies the containers to which it is applicable and conveys the information required by paragraph (f)(5) of this section to be on a label. The written materials shall be readily accessible to the employees in their work area throughout each work shift.

(7) The employer is not required to label portable containers into which hazardous chemicals are transferred from labeled containers, and which are intended only for the immediate use of the employee who performs the transfer. For purposes of this section, drugs which are dispensed by a pharmacy to a health care provider for direct administration to a patient are exempted from labeling.

(8) The employer shall not remove or deface existing labels on incoming containers of hazardous chemicals, unless the container is immediately marked with the required information.

(9) The employer shall ensure that labels or other forms of warning are legible, in English, and prominently displayed on the container, or readily available in the work area throughout each work shift. Employers having employees who speak other languages may add the information in their language to the material presented, as long as the information is presented in English as well.

(10) The chemical manufacturer, importer, distributor or employer need not affix new labels to comply with this section if existing labels already convey the required information.

(11) Chemical manufacturers, importers, distributors, or employers who become newly aware of any significant information regarding the hazards of a chemical shall revise the labels for the chemical within three months of becoming aware of the new information. Labels on containers of hazardous chemicals shipped after that time shall contain the new information. If the chemical is not currently produced or imported, the chemical manufacturer, importers, distributor, or employer shall add the information to the label before the chemical is shipped or introduced into the workplace again.

(g) "Material safety data sheets."

(1) Chemical manufacturers and importers shall obtain or develop a material safety data sheet for each hazardous chemical they produce or import. Employers shall have a material safety data sheet in the workplace for each hazardous chemical which they use.

(2) Each material safety data sheet shall be in English (although the employer may maintain copies in other languages as well), and shall contain at least the following information:

(i) The identity used on the label, and, except as provided for in paragraph (i) of this section on trade secrets:

(A) If the hazardous chemical is a single substance, its chemical and common name(s);

(B) If the hazardous chemical is a mixture which has been tested as a whole to determine its hazards, the chemical and common name(s) of the ingredients which contribute to these known hazards, and the common name(s) of the mixture itself; or,

(C) If the hazardous chemical is a mixture which has not been tested as a whole:

(1) The chemical and common name(s) of all ingredients which have been determined to be health hazards, and which comprise 1% or greater of the composition, except that chemicals identified as carcinogens under paragraph (d) of this section shall be listed if the concentrations are 0.1% or greater; and,

(2) The chemical and common name(s) of all ingredients which have been determined to be health hazards, and which comprise less than 1% (0.1% for carcinogens) of the mixture, if there is evidence that the ingredient(s) could be released from the mixture in concentrations which would exceed an established OSHA permissible exposure limit or ACGIH Threshold Limit Value, or could present a health risk to employees; and,

(3) The chemical and common name(s) of all ingredients which have been determined to present a physical hazard when present in the mixture;

(ii) Physical and chemical characteristics of the hazardous chemical (such as vapor pressure, flash point);

(iii) The physical hazards of the hazardous chemical, including the potential for fire, explosion, and reactivity;

(iv) The health hazards of the hazardous chemical, including signs and symptoms of exposure, and any medical conditions which are generally recognized as being aggravated by exposure to the chemical;

(v) The primary route(s) of entry;

(vi) The OSHA permissible exposure limit, ACGIH Threshold Limit Value, and any other exposure limit used or recommended by the chemical manufacturer, importer, or employer preparing the material safety data sheet, where available;

(vii) Whether the hazardous chemical is listed in the National Toxicology Program (NTP) Annual Report on Carcinogens (latest edition) or has been found to be a potential carcinogen in the International Agency for Research on Cancer (IARC) Monographs (latest editions), or by OSHA;

(viii) Any generally applicable precautions for safe handling and use which are known to the chemical manufacturer, importer or employer preparing the material safety data sheet, including appropriate hygienic practices, protective measures during repair and maintenance of contaminated equipment, and procedures for clean-up of spills and leaks;

(ix) Any generally applicable control measures which are known to the chemical manufacturer, importer or employer preparing the material safety data sheet, such as appropriate engineering controls, work practices, or personal protective equipment;

(x) Emergency and first aid procedures;

(xi) The date of preparation of the material safety data sheet or the last change to it; and,

(xii) The name, address and telephone number of the chemical manufacturer, importer, employer or other responsible party preparing or distributing the material safety data sheet, who can provide additional information on the hazardous chemical and appropriate emergency procedures, if necessary.

(3) If no relevant information is found for any given category on the material safety data sheet, the chemical manufacturer, importer or employer preparing the material safety data sheet shall mark it to indicate that no applicable information was found.

(4) Where complex mixtures have similar hazards and contents (i.e. the chemical ingredients are essentially the same, but the specific composition varies from mixture to mixture), the chemical manufacturer, importer or employer may prepare one material safety data sheet to apply to all of these similar mixtures.

(5) The chemical manufacturer, importer or employer preparing the material safety data sheet shall ensure that the information recorded accurately reflects the scientific evidence used in making the hazard determination. If the chemical manufacturer, importer or employer preparing the material safety data sheet becomes newly aware of any significant information regarding the hazards of a chemical, or ways to protect against the hazards, this new information shall be added to the material safety data sheet within three months. If the chemical is not currently being produced or imported the chemical manufacturer or importer shall add the information to the material safety data sheet before the chemical is introduced into the workplace again.

(6)(i) Chemical manufacturers or importers shall ensure that distributors and employers are provided an appropriate material safety data sheet with their initial shipment, and with the first shipment after a material safety data sheet is updated;

(ii) The chemical manufacturer or importer shall either provide material safety data sheets with the shipped containers or send them to the distributor or employer prior to or at the time of the shipment;

(iii) If the material safety data sheet is not provided with a shipment that has been labeled as a hazardous chemical, the distributor or employer shall obtain one from the chemical manufacturer or importer as soon as possible; and,

(iv) The chemical manufacturer or importer shall also provide distributors or employers with a material safety data sheet upon request.

(7)(i) Distributors shall ensure that material safety data sheets, and updated information, are provided to other distributors and employers with their initial shipment and with the first shipment after a material safety data sheet is updated;

(ii) The distributor shall either provide material safety data sheets with the shipped containers, or send them to the other distributor or employer prior to or at the time of the shipment;

(iii) Retail distributors selling hazardous chemicals to employers having a commercial account shall provide a material safety data sheet to such employers upon request, and shall post a sign or otherwise inform them that a material safety data sheet is available;

(iv) Wholesale distributors selling hazardous chemicals to employers over-the-counter may also provide material safety data sheets upon the request of the employer at the time of the over-the-counter purchase, and shall post a sign or otherwise inform such employers that a material safety data sheet is available;

(v) If an employer without a commercial account purchases a hazardous chemical from a retail distributor not required to have material safety data sheets on file (i.e., the retail distributor does not have commercial accounts and does not use the materials), the retail distributor shall provide the employer, upon request, with the name, address, and telephone number of the chemical manufacturer, importer, or distributor from which a material safety data sheet can be obtained;

(vi) Wholesale distributors shall also provide material safety data sheets to employers or other distributors upon request; and,

(vii) Chemical manufacturers, importers, and distributors need not provide material safety data sheets to retail distributors that have informed them that the retail distributor does not sell the product to commercial accounts or open the sealed container to use it in their own workplaces.

(8) The employer shall maintain in the workplace copies of the required material safety data sheets for each hazardous chemical, and shall ensure that they are readily accessible during each work shift to employees when they are in their work area(s). (Electronic access, microfiche, and other alternatives to maintaining paper copies of the material safety data sheets are permitted as long as no barriers to immediate employee access in each workplace are created by such options.)

(9) Where employees must travel between workplaces during a workshift, i.e., their work is carried out at more than one geographical location, the material safety data sheets may be kept at the primary workplace facility. In this situation, the employer shall ensure that employees can immediately obtain the required information in an emergency.

(10) Material safety data sheets may be kept in any form, including operating procedures, and may be designed to cover groups of hazardous chemicals in a work area where it may be more appropriate to address the hazards of a process rather than individual hazardous chemicals. However, the employer shall ensure that in all cases the required information is provided for each hazardous chemical, and is readily accessible during each work shift to employees when they are in in their work area(s).

(11) Material safety data sheets shall also be made readily available, upon request, to designated representatives and to the Assistant Secretary, in accordance with the requirements of 29 CFR 1910.1020(e). The Director shall also be given access to material safety data sheets in the same manner.

(h) "Employee information and training." (1) Employers shall provide employees with effective information and training on hazardous chemicals in their work area at the time of their initial assignment, and whenever a new physical or health hazard the employees have not previously been trained about is introduced into their work area. Information and training may be designed to cover categories of hazards (e.g., flammability, carcinogenicity) or specific chemicals. Chemical-specific information must always be available through labels and material safety data sheets.

(2) "Information." Employees shall be informed of:

(i) The requirements of this section;

(ii) Any operations in their work area where hazardous chemicals are present; and,

(iii) The location and availability of the written hazard communication program, including the required list(s) of hazardous chemicals, and material safety data sheets required by this section.

1910.1200(h)(3) Content of Employee Training - Training Required

(3) "Training." Employee training shall include at least:

(i) Methods and observations that may be used to detect the presence or release of a hazardous chemical in the work area (such as monitoring conducted by the employer, continuous monitoring devices, visual appearance or odor of hazardous chemicals when being released, etc.);

(ii) The physical and health hazards of the chemicals in the work area;

(iii) The measures employees can take to protect themselves from these hazards, including specific procedures the employer has implemented to protect employees from exposure to hazardous chemicals, such as appropriate work practices, emergency procedures, and personal protective equipment to be used; and,

(iv) The details of the hazard communication program developed by the employer, including an explanation of the labeling system and the material safety data sheet, and how employees can obtain and use the appropriate hazard information.

(i) "Trade secrets."

(1) The chemical manufacturer, importer, or employer may withhold the specific chemical identity, including the chemical name and other specific identification of a hazardous chemical, from the material safety data sheet, provided that:

(i) The claim that the information withheld is a trade secret can be supported;

(ii) Information contained in the material safety data sheet concerning the properties and effects of the hazardous chemical is disclosed;

(iii) The material safety data sheet indicates that the specific chemical identity is being withheld as a trade secret; and,

(iv) The specific chemical identity is made available to health professionals, employees, and designated representatives in accordance with the applicable provisions of this paragraph.

(2) Where a treating physician or nurse determines that a medical emergency exists and the specific chemical identity of a hazardous chemical is necessary for emergency or first-aid treatment, the chemical manufacturer, importer, or employer shall immediately disclose the specific chemical identity of a trade secret chemical to that treating physician or nurse, regardless of the existence of a written statement of need or a confidentiality agreement. The chemical manufacturer, importer, or employer may require a written statement of need and confidentiality agreement, in accordance with the provisions of paragraphs (i)(3) and (4) of this section, as soon as circumstances permit.

(3) In non-emergency situations, a chemical manufacturer, importer, or employer shall, upon request, disclose a specific chemical identity, otherwise permitted to be withheld under paragraph (i)(1) of this section, to a health professional (i.e. physician, industrial hygienist, toxicologist, epidemiologist, or occupational health nurse) providing medical or other occupational health services to exposed employee(s), and to employees or designated representatives, if:

(i) The request is in writing;

(ii) The request describes with reasonable detail one or more of the following occupational health needs for the information:

(A) To assess the hazards of the chemicals to which employees will be exposed;

(B) To conduct or assess sampling of the workplace atmosphere to determine employee exposure levels;

(C) To conduct pre-assignment or periodic medical surveillance of exposed employees;

(D) To provide medical treatment to exposed employees;

(E) To select or assess appropriate personal protective equipment for exposed employees;

(F) To design or assess engineering controls or other protective measures for exposed employees; and,

(G) To conduct studies to determine the health effects of exposure.

(iii) The request explains in detail why the disclosure of the specific chemical identity is essential and that, in lieu thereof, the disclosure of the following information to the health professional, employee, or designated representative, would not satisfy the purposes described in paragraph (i)(3)(ii) of this section:

(A) The properties and effects of the chemical;

(B) Measures for controlling workers' exposure to the chemical;

(C) Methods of monitoring and analyzing worker exposure to the chemical; and,

(D) Methods of diagnosing and treating harmful exposures to the chemical;

(iv) The request includes a description of the procedures to be used to maintain the confidentiality of the disclosed information; and,

(v) The health professional, and the employer or contractor of the services of the health professional (i.e. downstream employer, labor organization, or individual employee), employee, or designated representative, agree in a written confidentiality agreement that the health professional, employee, or designated representative, will not use the trade secret information for any purpose other than the health need(s) asserted and agree not to release the information under any circumstances other than to OSHA, as provided in paragraph (i)(6) of this section, except as authorized by the terms of the agreement or by the chemical manufacturer, importer, or employer.

(4) The confidentiality agreement authorized by paragraph (i)(3)(iv) of this section:

(i) May restrict the use of the information to the health purposes indicated in the written statement of need;

(ii) May provide for appropriate legal remedies in the event of a breach of the agreement, including stipulation of a reasonable pre-estimate of likely damages; and,

(iii) May not include requirements for the posting of a penalty bond.

(5) Nothing in this standard is meant to preclude the parties from pursuing non-contractual remedies to the extent permitted by law.

(6) If the health professional, employee, or designated representative receiving the trade secret information decides that there is a need to disclose it to OSHA, the chemical manufacturer, importer, or employer who provided the information shall be informed by the health professional, employee, or designated representative prior to, or at the same time as, such disclosure.

(7) If the chemical manufacturer, importer, or employer denies a written request for disclosure of a specific chemical identity, the denial must:

(i) Be provided to the health professional, employee, or designated representative, within thirty days of the request;

(ii) Be in writing;

(iii) Include evidence to support the claim that the specific chemical identity is a trade secret;

(iv) State the specific reasons why the request is being denied; and,

(v) Explain in detail how alternative information may satisfy the specific medical or occupational health need without revealing the specific chemical identity.

(8) The health professional, employee, or designated representative whose request for information is denied under paragraph (i)(3) of this section may refer the request and the written denial of the request to OSHA for consideration.

(9) When a health professional, employee, or designated representative refers the denial to OSHA under paragraph (i)(8) of this section, OSHA shall consider the evidence to determine if:

(i) The chemical manufacturer, importer, or employer has supported the claim that the specific chemical identity is a trade secret;

(ii) The health professional, employee, or designated representative has supported the claim that there is a medical or occupational health need for the information; and,

(iii) The health professional, employee or designated representative has demonstrated adequate means to protect the confidentiality.

(10)(i) If OSHA determines that the specific chemical identity requested under paragraph (i)(3) of this section is not a "bona fide" trade secret, or that it is a trade secret, but the requesting health professional, employee, or designated representative has a legitimate medical or occupational health need for the information, has executed a written confidentiality agreement, and has shown adequate means to protect the confidentiality of the information, the chemical manufacturer, importer, or employer will be subject to citation by OSHA.

(ii) If a chemical manufacturer, importer, or employer demonstrates to OSHA that the execution of a confidentiality agreement would not provide sufficient protection against the potential harm from the unauthorized disclosure of a trade secret specific chemical identity, the Assistant Secretary may issue such orders or impose such additional limitations or conditions upon the disclosure of the requested chemical information as may be appropriate to assure that the occupational health services are provided without an undue risk of harm to the chemical manufacturer, importer, or employer.

(11) If a citation for a failure to release specific chemical identity information is contested by the chemical manufacturer, importer, or employer, the matter will be adjudicated before the Occupational Safety and Health Review Commission in accordance with the Act's enforcement scheme and the applicable Commission rules of procedure. In accordance with the Commission rules, when a chemical manufacturer, importer, or employer continues to withhold the information during the contest, the Administrative Law Judge may review the citation and supporting documentation "in camera" or issue appropriate orders to protect the confidentiality of such matters.

(12) Notwithstanding the existence of a trade secret claim, a chemical manufacturer, importer, or employer shall, upon request, disclose to the Assistant Secretary any information which this section requires the chemical manufacturer, importer, or employer to make available. Where there is a trade secret claim, such claim shall be made no later than at the time the information is provided to the Assistant Secretary so that suitable determinations of trade secret status can be made and the necessary protections can be implemented.

(13) Nothing in this paragraph shall be construed as requiring the disclosure under any circumstances of process or percentage of mixture information which is a trade secret.

(j) "Effective dates." Chemical manufacturers, importers, distributors, and employers shall be in compliance with all provisions of this section by March 11, 1994.

*Note:* *The effective date of the clarification that the exemption of wood and wood products from the Hazard Communication standard in paragraph (b)(6)(iv) only applies to wood and wood products including lumber which will not be processed, where the manufacturer or importer can establish that the only hazard they pose to employees is the potential for flammability or combustibility, and that the exemption does not apply to wood or wood products which have been treated with a hazardous chemical covered by this standard, and wood which may be subsequently sawed or cut generating dust has been stayed from March 11, 1994 to August 11, 1994.*

# 1910.1200, Appendix A - Health Hazard Definitions (Mandatory)

Although safety hazards related to the physical characteristics of a chemical can be objectively defined in terms of testing requirements (e.g. flammability), health hazard definitions are less precise and more subjective. Health hazards may cause measurable changes in the body - such as decreased pulmonary function. These changes are generally indicated by the occurrence of signs and symptoms in the exposed employees - such as shortness of breath, a non-measurable, subjective feeling. Employees exposed to such hazards must be apprised of both the change in body function and the signs and symptoms that may occur to signal that change.

The determination of occupational health hazards is complicated by the fact that many of the effects or signs and symptoms occur commonly in non-occupationally exposed populations, so that effects of exposure are difficult to separate from normally occurring illnesses. Occasionally, a substance causes an effect that is rarely seen in the population at large, such as angiosarcomas caused by vinyl chloride exposure, thus making it easier to ascertain that the occupational exposure was the primary causative factor. More often, however, the effects are common, such as lung cancer. The situation is further complicated by the fact that most chemicals have not been adequately tested to determine their health hazard potential, and data do not exist to substantiate these effects.

There have been many attempts to categorize effects and to define them in various ways. Generally, the terms "acute" and "chronic" are used to delineate between effects on the basis of severity or duration. "Acute" effects usually occur rapidly as a result of short-term exposures, and are of short duration. "Chronic" effects generally occur as a result of long-term exposure, and are of long duration.

The acute effects referred to most frequently are those defined by the American National Standards Institute (ANSI) standard for Precautionary Labeling of Hazardous Industrial Chemicals (Z129.1-1988) - irritation, corrosivity, sensitization and lethal dose. Although these are important health effects, they do not adequately cover the considerable range of acute effects which may occur as a result of occupational exposure, such as, for example, narcosis.

Similarly, the term chronic effect is often used to cover only carcinogenicity, teratogenicity, and mutagenicity. These effects are obviously a concern in the workplace, but again, do not adequately cover the area of chronic effects, excluding, for example, blood dyscrasias (such as anemia), chronic bronchitis, and liver atrophy.

The goal of defining precisely, in measurable terms, every possible health effect that may occur in the workplace as a result of chemical exposures cannot realistically be accomplished. This does not negate the need for employees to be informed of such effects and protected from them.

Appendix B, which is also mandatory, outlines the principles and procedures of hazard assessment.

For purposes of this section, any chemicals which meet any of the following definitions, as determined by the criteria set forth in Appendix B, are health hazards. However, this is not intended to be an exclusive categorization scheme. If there are available scientific data that involve other animal species or test methods, they must also be evaluated to determine the applicability of the Hazard Communication Standard.

1. "Carcinogen:" A chemical is considered to be a carcinogen if:

   (a) It has been evaluated by the International Agency for Research on Cancer (IARC), and found to be a carcinogen or potential carcinogen; or

   (b) It is listed as a carcinogen or potential carcinogen in the Annual Report on Carcinogens published by the National Toxicology Program (NTP) (latest edition); or,

   (c) It is regulated by OSHA as a carcinogen.

2. "Corrosive:" A chemical that causes visible destruction of, or irreversible alterations in, living tissue by chemical action at the site of contact. For example, a chemical is considered to be corrosive if, when tested on the intact skin of albino rabbits by the method described by the U.S. Department of Transportation in Appendix A to

49 CFR Part 173, it destroys or changes irreversibly the structure of the tissue at the site of contact following an exposure period of four hours. This term shall not refer to action on inanimate surfaces.

3. "Highly toxic:" A chemical falling within any of the following categories:

(a) A chemical that has a median lethal dose (LD(50)) of 50 milligrams or less per kilogram of body weight when administered orally to albino rats weighing between two and three kilograms each.

(b) A chemical that has a median lethal dose (LD(50)) of 200 milligrams or less per kilogram of body weight when administered by continuous contact for 24 hours (or less if death occurs within 24 hours) with the bare skin of albino rabbits weighing between two and three kilograms each.

(c) A chemical that has a median lethal concentration (LC(50)) in air of 200 parts per million by volume or less of gas or vapor, or 2 milligrams per liter or less of mist, fume, or dust, when administered by continuous inhalation for one hour (or less if death occurs within one hour) to albino rats weighing between 200 and 300 grams each.

4. "Irritant:" A chemical, which is not corrosive, but which causes a reversible inflammatory effect on living tissue by chemical action at the site of contact. A chemical is a skin irritant if, when tested on the intact skin of albino rabbits by the methods of 16 CFR 1500.41 for four hours exposure or by other appropriate techniques, it results in an empirical score of five or more. A chemical is an eye irritant if so determined under the procedure listed in 16 CFR 1500.42 or other appropriate techniques.

5. "Sensitizer:" A chemical that causes a substantial proportion of exposed people or animals to develop an allergic reaction in normal tissue after repeated exposure to the chemical.

6. "Toxic." A chemical falling within any of the following categories:

(a) A chemical that has a median lethal dose (LD(50)) of more than 50 milligrams per kilogram but not more than 500 milligrams per kilogram of body weight when administered orally to albino rats weighing between two and three kilograms each.

(b) A chemical that has a median lethal dose (LD(50)) of more than 200 milligrams per kilogram but not more than 1,000 milligrams per kilogram of body weight when administered by continuous contact for 24 hours (or less if death occurs within 24 hours) with the bare skin of albino rabbits weighing between two and three kilograms each.

(c) A chemical that has a median lethal concentration (LC(50)) in air of more than 200 parts per million but not more than 2,000 parts per million by volume of gas or vapor, or more than two milligrams per liter but not more than 20 milligrams per liter of mist, fume, or dust, when administered by continuous inhalation for one hour (or less if death occurs within one hour) to albino rats weighing between 200 and 300 grams each.

7. "Target organ effects." The following is a target organ categorization of effects which may occur, including examples of signs and symptoms and chemicals which have been found to cause such effects. These examples are presented to illustrate the range and diversity of effects and hazards found in the workplace, and the broad scope employers must consider in this area, but are not intended to be all-inclusive.

a. Hepatotoxins: Chemicals which produce liver damage

Signs & Symptoms: Jaundice; liver enlargement

Chemicals: Carbon tetrachloride; nitrosamines

b. Nephrotoxins: Chemicals which produce kidney damage

Signs & Symptoms: Edema; proteinuria

Chemicals: Halogenated hydrocarbons; uranium

c. Neurotoxins: Chemicals which produce their primary toxic effects on the nervous system

Signs & Symptoms: Narcosis; behavioral changes; decrease in motor functions

Chemicals: Mercury; carbon disulfide

d. Agents which act on the blood or hemato-poietic system: Decrease hemoglobin function; deprive the body tissues of oxygen

Signs & Symptoms: Cyanosis; loss of consciousness

Chemicals: Carbon monoxide; cyanides

e. Agents which damage the lung: Chemicals which irritate or damage pulmonary tissue

Signs & Symptoms: Cough; tightness in chest; shortness of breath

Chemicals: Silica; asbestos

f. Reproductive toxins: Chemicals which affect the reproductive capabilities including chromosomal damage (mutations) and effects on fetuses (teratogenesis)

Signs & Symptoms: Birth defects; sterility

Chemicals: Lead; DBCP

g. Cutaneous hazards: Chemicals which affect the dermal layer of the body

Signs & Symptoms: Defatting of the skin; rashes; irritation

Chemicals: Ketones; chlorinated compounds

h. Eye hazards: Chemicals which affect the eye or visual capacity

Signs & Symptoms: Conjunctivitis; corneal damage

Chemicals: Organic solvents; acids

# 1910.1200, Appendix B - Hazard Determination (Mandatory)

The quality of a hazard communication program is largely dependent upon the adequacy and accuracy of the hazard determination. The hazard determination requirement of this standard is performance-oriented.

Chemical manufacturers, importers, and employers evaluating chemicals are not required to follow any specific methods for determining hazards, but they must be able to demonstrate that they have adequately ascertained the hazards of the chemicals produced or imported in accordance with the criteria set forth in this Appendix.

Hazard evaluation is a process which relies heavily on the professional judgment of the evaluator, particularly in the area of chronic hazards.

The performance-orientation of the hazard determination does not diminish the duty of the chemical manufacturer, importer or employer to conduct a thorough evaluation, examining all relevant data and producing a scientifically defensible evaluation. For purposes of this standard, the following criteria shall be used in making hazard determinations that meet the requirements of this standard.

1. "Carcinogenicity:" As described in paragraph (d)(4) of this section and Appendix A of this section, a determination by the National Toxicology Program, the International Agency for Research on Cancer, or OSHA that a chemical is a carcinogen or potential carcinogen will be considered conclusive evidence for purposes of this section. In addition, however, all available scientific data on carcinogenicity must be evaluated in accordance with the provisions of this Appendix and the requirements of the rule.

2. "Human data:" Where available, epidemiological studies and case reports of adverse health effects shall be considered in the evaluation.

3. "Animal data:" Human evidence of health effects in exposed populations is generally not available for the majority of chemicals produced or used in the workplace. Therefore, the available results of toxicological testing in animal populations shall be used to predict the health effects that may be experienced by exposed workers. In particular, the definitions of certain acute hazards refer to specific animal testing results (see Appendix A).

4. "Adequacy and reporting of data." The results of any studies which are designed and conducted according to established scientific principles, and which report statistically significant conclusions regarding the health effects of a chemical, shall be a sufficient basis for a hazard determination and reported on any material safety data sheet. In vitro studies alone generally do not form the basis for a definitive finding of hazard under the HCS since they have a positive or negative result rather than a statistically significant finding.

The chemical manufacturer, importer, or employer may also report the results of other scientifically valid studies which tend to refute the findings of hazard.

# 1910.1200, Appendix C - Information Sources (Advisory)

Editorial Note: Removed by the Federal Register as of March 7, 1996.

# 1910.1200, Appendix D - Definition of "Trade Secret" (Mandatory)

The following is a reprint of the "Restatement of Torts" Section 757, Comment B (1939):

b. "Definition of trade secret." A trade secret may consist of any formula, pattern, device or compilation of information which is used in one's business, and which gives him an opportunity to obtain an advantage over competitors who do not know or use it. It may be a formula for a chemical compound, a process of manufacturing, treating or preserving materials, a pattern for a machine or other device, or a list of customers. It differs from other secret information in a business (see s759 of the Restatement of Torts which is not included in this Appendix) in that it is not simply information as to single or ephemeral events in the conduct of the business, as, for example, the amount or other terms of a secret bid for a contract or the salary of certain employees, or the security investments made or contemplated, or the date fixed for the announcement of a new policy or for bringing out a new model or the like. A trade secret is a process or device for continuous use in the operations of the business. Generally it relates to the production of goods, as, for example, a machine or formula for the production of an article. It may, however, relate to the sale of goods or to other operations in the business, such as a code for determining discounts, rebates or other concessions in a price list or catalogue, or a list of specialized customers, or a method of bookkeeping or other office management.

"Secrecy." The subject matter of a trade secret must be secret. Matters of public knowledge or of general knowledge in an industry cannot be appropriated by one as his secret. Matters which are completely disclosed by the goods which one markets cannot be his secret. Substantially, a trade secret is known only in the particular business in which it is used. It is not requisite that only the proprietor of the business know it. He may, without losing his protection, communicate it to employees involved in its use. He may likewise communicate it to others pledged to secrecy. Others may also know of it independently, as, for example, when they have discovered the process or formula by independent invention and are keeping it secret. Nevertheless, a substantial element of secrecy must exist, so that, except by the use of improper means, there would be difficulty in acquiring the information. An exact definition of a trade secret is not possible. Some factors to be considered in determining whether given information is one's trade secret are: (1) The extent to which the information is known outside of his business; (2) the extent to which it is known by employees and others involved in his business; (3) the extent of measures taken by him to guard the secrecy of the information; (4) the value of the information to him and his competitors; (5) the amount of effort or money expended by him in developing the information; (6) the ease or difficulty with which the information could be properly acquired or duplicated by others.

"Novelty and prior art." A trade secret may be a device or process which is patentable; but it need not be that. It may be a device or process which is clearly anticipated in the prior art or one which is merely a mechanical improvement that a good mechanic can make. Novelty and invention are not requisite for a trade secret as they are for patentability. These requirements are essential to patentability because a patent protects against unlicensed use of the patented device or process even by one who discovers it properly through independent research. The patent monopoly is a reward to the inventor. But such is not the case with a trade secret. Its protection is not based on a policy of rewarding or otherwise encouraging the development of secret processes or devices. The protection is merely against breach of faith and reprehensible means of learning another's secret. For this limited protection it is not appropriate to require also the kind of novelty and invention which is a requisite of patentability. The nature of the secret is, however, an important factor in determining the kind of relief that is appropriate against one who is subject to liability under the rule stated in this Section. Thus, if the secret consists of a device or process which is a novel invention, one who acquires the secret wrongfully is ordinarily enjoined from further use of it and is required to account for the profits derived from his past use. If, on the other hand, the secret consists of mechanical improvements that a good mechanic can make without resort to the secret, the wrongdoer's liability may be limited to damages, and an injunction against future use of the improvements made with the aid of the secret may be inappropriate.

# 1910.1200, Appendix E - Guidelines for Employer Compliance (Advisory)

The Hazard Communication Standard (HCS) is based on a simple concept – that employees have both a need and a right to know the hazards and identities of the chemicals they are exposed to when working. They also need to know what protective measures are available to prevent adverse effects from occurring. The HCS is designed to provide employees with the information they need.

Knowledge acquired under the HCS will help employers provide safer workplaces for their employees. When employers have information about the chemicals being used, they can take steps to reduce exposures, substitute less hazardous materials, and establish proper work practices. These efforts will help prevent the occurrence of work-related illnesses and injuries caused by chemicals.

The HCS addresses the issues of evaluating and communicating hazards to workers. Evaluation of chemical hazards involves a number of technical concepts, and is a process that requires the professional judgment of experienced experts. That's why the HCS is designed so that employers who simply use chemicals, rather than produce or import them, are not required to evaluate the hazards of those chemicals. Hazard determination is the responsibility of the producers and importers of the materials. Producers and importers of chemicals are then required to provide the hazard information to employers that purchase their products.

Employers that don't produce or import chemicals need only focus on those parts of the rule that deal with establishing a workplace program and communicating information to their workers. This appendix is a general guide for such employers to help them determine what's required under the rule. It does not supplant or substitute for the regulatory provisions, but rather provides a simplified outline of the steps an average employer would follow to meet those requirements.

1. "Becoming Familiar With The Rule"

OSHA has provided a simple summary of the HCS in a pamphlet entitled "Chemical Hazard Communication," OSHA Publication Number 3084. Some employers prefer to begin to become familiar with the rule's requirements by reading this pamphlet. A copy may be obtained from your local OSHA Area Office, or by contacting the OSHA Publications Office at (202) 523-9667.

The standard is long, and some parts of it are technical, but the basic concepts are simple. In fact, the requirements reflect what many employers have been doing for years. You may find that you are already largely in compliance with many of the provisions, and will simply have to modify your existing programs somewhat. If you are operating in an OSHA-approved State Plan State, you must comply with the state's requirements, which may be different than those of the Federal rule. Many of the State Plan States had hazard communication or "right-to-know" laws prior to promulgation of the Federal rule. Employers in State Plan States should contact their State OSHA offices for more information regarding applicable requirements.

The HCS requires information to be prepared and transmitted regarding all hazardous chemicals. The HCS covers both physical hazards (such as flammability), and health hazards (such as irritation, lung damage, and cancer). Most chemicals used in the workplace have some hazard potential, and thus will be covered by the rule.

One difference between this rule and many others adopted by OSHA is that this one is performance-oriented. That means that you have the flexibility to adapt the rule to the needs of your workplace, rather than having to follow specific, rigid requirements. It also means that you have to exercise more judgment to implement an appropriate and effective program.

The standard's design is simple. Chemical manufacturers and importers must evaluate the hazards of the chemicals they produce or import. Using that information, they must then prepare labels for containers, and more detailed technical bulletins called material safety data sheets (MSDS).

Chemical manufacturers, importers, and distributors of hazardous chemicals are all required to provide the appropriate labels and material safety data sheets to the employers to which they ship the chemicals. The information is to be provided automatically. Every container of hazardous chemicals you receive must be labeled, tagged, or marked with the required information. Your suppliers must also send you a properly completed MSDS at the time of the first shipment of the chemical, and with the next shipment after the MSDS is updated with new and significant information about the hazards.

You can rely on the information received from your suppliers. You have no independent duty to analyze the chemical or evaluate the hazards of it.

Employers that "use" hazardous chemicals must have a program to ensure the information is provided to exposed employees. "Use" means to package, handle, react, or transfer. This is an intentionally broad scope, and includes any situation where a chemical is present in such a way that employees may be exposed under normal conditions of use or in a foreseeable emergency.

The requirements of the rule that deal specifically with the hazard communication program are found in this section in paragraphs (e), written hazard communication program; (f), labels and other forms of warning; (g), material safety data sheets; and (h), employee information and training. The requirements of these paragraphs should be the focus of your attention. Concentrate on becoming familiar with them, using paragraphs (b), scope and application, and (c), definitions, as references when needed to help explain the provisions.

There are two types of work operations where the coverage of the rule is limited. These are laboratories and operations where chemicals are only handled in sealed containers (e.g., a warehouse). The limited provisions for these workplaces can be found in Paragraph (b) of this section, scope and application. Basically, employers having these types of work operations need only keep labels on containers as they are received; maintain material safety data sheets that are received, and give employees access to them; and provide information and training for employees. Employers do not have to have written hazard communication programs and lists of chemicals for these types of operations.

The limited coverage of laboratories and sealed container operations addresses the obligation of an employer to the workers in the operations involved, and does not affect the employer's duties as a distributor of chemicals. For example, a distributor may have warehouse operations where employees would be protected under the limited sealed container provisions. In this situation, requirements for obtaining and maintaining MSDSs are limited to providing access to those received with containers while the substance is in the workplace, and requesting MSDSs when employees request access for those not received with the containers. However, as a distributor of hazardous chemicals, that employer will still have responsibilities for providing MSDSs to downstream customers at the time of the first shipment and when the MSDS is updated. Therefore, although they may not be required for the employees in the work operation, the distributor may, nevertheless, have to have MSDSs to satisfy other requirements of the rule.

2. "Identify Responsible Staff"

Hazard communication is going to be a continuing program in your facility. Compliance with the HCS is not a "one shot deal." In order to have a successful program, it will be necessary to assign responsibility for both the initial and ongoing activities that have to be undertaken to comply with the rule. In some cases, these activities may already be part of current job assignments. For example, site supervisors are frequently responsible for on-the-job training sessions. Early identification of the responsible employees, and involvement of them in the development of your plan of action, will result in a more effective program design. Evaluation of the effectiveness of your program will also be enhanced by involvement of affected employees.

For any safety and health program, success depends on commitment at every level of the organization. This is particularly true for hazard communication, where success requires a change in behavior. This will only occur if employers understand the program, and are committed to its success, and if employees are motivated by the people presenting the information to them.

3. "Identify Hazardous Chemicals in the Workplace"

The standard requires a list of hazardous chemicals in the workplace as part of the written hazard communication program. The list will eventually serve as an inventory of everything for which an MSDS must be maintained. At this point, however, preparing the list will help you complete the rest of the program since it will give you some idea of the scope of the program required for compliance in your facility.

The best way to prepare a comprehensive list is to survey the workplace. Purchasing records may also help, and certainly employers should establish procedures to ensure that in the future purchasing procedures result in MSDSs being received before a material is used in the workplace.

The broadest possible perspective should be taken when doing the survey. Sometimes people think of "chemicals" as being only liquids in containers. The HCS covers chemicals in all physical forms – liquids, solids, gases, vapors, fumes, and mists – whether they are "contained" or not. The hazardous nature of the chemical and the potential for exposure are the factors which determine whether a chemical is covered. If it's not hazardous, it's not covered. If there is no potential for exposure (e.g., the chemical is inextricably bound and cannot be released), the rule does not cover the chemical.

Look around. Identify chemicals in containers, including pipes, but also think about chemicals generated in the work operations. For example, welding fumes, dusts, and exhaust fumes are all sources of chemical exposures. Read labels provided by suppliers for hazard information. Make a list of all chemicals in the workplace that are potentially hazardous. For your own information and planning, you may also want to note the location(s) of the products within the workplace, and an indication of the hazards as found on the label. This will help as you prepare the rest of your program.

Paragraph (b) of this section, scope and application, includes exemptions for various chemicals or workplace situations. After compiling the complete list of chemicals, you should review paragraph (b) of this section to determine if any of the items can be eliminated from the list because they are exempted materials. For example, food, drugs, and cosmetics brought into the workplace for employee consumption are exempt. So rubbing alcohol in the first aid kit would not be covered.

Once you have compiled as complete a list as possible of the potentially hazardous chemicals in the workplace, the next step is to determine if you have received material safety data sheets for all of them. Check your files against the inventory you have just compiled. If any are missing, contact your supplier and request one. It is a good idea to document these requests, either by copy of a letter or a note regarding telephone conversations. If you have MSDSs for chemicals that are not on your list, find out why. Maybe you don't use the chemical anymore. Or maybe you missed it in your survey. Some suppliers do provide MSDSs for products that are not hazardous. These do not have to be maintained by you.

You should not allow employees to use any chemicals for which you have not received an MSDS. The MSDS provides information you need to ensure proper protective measures are implemented prior to exposure.

4. "Preparing and Implementing a Hazard Communication Program"

All workplaces where employees are exposed to hazardous chemicals must have a written plan which describes how the standard will be implemented in that facility. Preparation of a plan is not just a paper exercise – all of the elements must be implemented in the workplace in order to be in compliance with the rule. See Paragraph (e) of this section for the specific requirements regarding written hazard communication programs. The only work operations which do not have to comply with the written plan requirements are laboratories and work operations where employees only handle chemicals in sealed containers. See Paragraph (b) of this section, scope and application, for the specific requirements for these two types of workplaces.

The plan does not have to be lengthy or complicated. It is intended to be a blueprint for implementation of your program – an assurance that all aspects of the requirements have been addressed.

Many trade associations and other professional groups have provided sample programs and other assistance materials to affected employers. These have been very helpful to many employers since they tend to be tailored to the particular industry involved. You may wish to investigate whether your industry trade groups have developed such materials.

Although such general guidance may be helpful, you must remember that the written program has to reflect what you are doing in your workplace. Therefore, if you use a generic program it must be adapted to address the facility it covers. For example, the written plan must list the chemicals present at the site, indicate who is to be responsible for the various aspects of the program in your facility, and indicate where written materials will be made available to employees.

If OSHA inspects your workplace for compliance with the HCS, the OSHA compliance officer will ask to see your written plan at the outset of the inspection. In general, the following items will be considered in evaluating your program.

The written program must describe how the requirements for labels and other forms of warning, material safety data sheets, and employee information and training, are going to be met in your facility. The following discussion provides the type of information compliance officers will be looking for to decide whether these elements of the hazard communication program have been properly addressed:

A. "Labels and Other Forms of Warning"

In-plant containers of hazardous chemicals must be labeled, tagged, or marked with the identity of the material and appropriate hazard warnings. Chemical manufacturers, importers, and distributors are required to ensure that every container of hazardous chemicals they ship is appropriately labeled with such information and with the name and address of the producer or other responsible party. Employers purchasing chemicals can rely on the labels provided by their suppliers. If the material is subsequently transferred by the employer from a labeled container to another container, the employer will have to label that container unless it is subject to the portable container exemption. See Paragraph (f) of this section for specific labeling requirements.

The primary information to be obtained from an OSHA-required label is the identity of the material and appropriate hazard warnings. The identity is any term which appears on the label, the MSDS, and the list of chemicals, and thus links these three sources of information. The identity used by the supplier may be a common or trade name ("Black Magic Formula"), or a chemical name (1,1,1,-trichloroethane). The hazard warning is a brief statement of the hazardous effects of the chemical ("flammable," "causes lung damage"). Labels frequently contain other information, such as precautionary measures ("do not use near open flame"), but this information is provided voluntarily and is not required by the rule. Labels must be legible, and prominently displayed. There are no specific requirements for size or color, or any specified text.

With these requirements in mind, the compliance officer will be looking for the following types of information to ensure that labeling will be properly implemented in your facility:

1. Designation of person(s) responsible for ensuring labeling of in-plant containers;
2. Designation of person(s) responsible for ensuring labeling of any shipped containers;
3. Description of labeling system(s) used;
4. Description of written alternatives to labeling of in-plant containers (if used); and,
5. Procedures to review and update label information when necessary.

Companies that are purchasing and using hazardous chemicals – rather than producing or distributing them – will primarily be concerned with ensuring that every purchased container is labeled. If materials are transferred into other containers, the employer must ensure that these are labeled as well, unless they fall under the portable container exemption (Paragraph (f)(7) of this section). In terms of labeling systems, you can simply choose to use the labels provided by your suppliers on the containers. These will generally be verbal text labels, and do not usually include numerical rating systems or symbols that require special training. The most important thing to remember is that this is a continuing duty – all in-plant containers of hazardous chemicals must always be labeled. Therefore, it is important to designate someone to be responsible for ensuring that the labels are maintained as required on the containers in your facility, and that newly purchased materials are checked for labels prior to use.

B. "Material Safety Data Sheets"

Chemical manufacturers and importers are required to obtain or develop an MSDS for each hazardous chemical they produce or import. Distributors are responsible for ensuring that their customers are provided a copy of these MSDSs. Employers must have an MSDS for each hazardous chemical which they use. Employers may rely on the information received from their suppliers. The specific requirements for MSDSs are in Paragraph (g) of this section. There is no specified format for the MSDS under the rule, although there are specific information requirements. OSHA has developed a non-mandatory format, OSHA Form 174, which may be used by chemical manufacturers and importers to comply with the rule. The MSDS must be in English. You are entitled to receive from your supplier a data sheet which includes all of the information required under the rule. If you do not receive one automatically, you should request one. If you receive one that is obviously inadequate, with, for example, blank spaces that are not completed, you should request an appropriately completed one. If your request for a data sheet or for a corrected data sheet does not produce the information needed, you should contact your local OSHA Area Office for assistance in obtaining the MSDS.

The role of MSDSs under the rule is to provide detailed information on each hazardous chemical, including its potential hazardous effects, its physical and chemical characteristics, and recommendations for appropriate protective measures. This information should be useful to you as the employer responsible for designing protective programs, as well as to the workers. If you are not familiar with MSDSs and with chemical terminology, you may need to learn to use them yourself. A glossary of MSDS terms may be helpful in this regard. Generally speaking, most employers using hazardous chemicals will primarily be concerned with MSDS information regarding hazardous effects and recommended protective measures. Focus on the sections of the MSDS that are applicable to your situation.

MSDSs must be readily accessible to employees when they are in their work areas during their workshifts. This may be accomplished in many different ways. You must decide what is appropriate for your particular workplace. Some employers keep the MSDSs in a binder in a central location (e.g., in the pick-up truck on a construction site). Others, particularly in workplaces with large numbers of chemicals, computerize the information and provide access through terminals. As long as employees can get the information when they need it, any approach may be used. The employees must have access to the MSDSs themselves – simply having a system where the information can be read to them over the phone is only permitted under the mobile worksite provision, Paragraph (g)(9) of this section, when employees must travel between workplaces during the shift. In this situation, they have access to the MSDSs prior to leaving the primary worksite, and when they return, so the telephone system is simply an emergency arrangement.

In order to ensure that you have a current MSDS for each chemical in the plant as required, and that employee access is provided, the compliance officers will be looking for the following types of information in your written program:

1. Designation of person(s) responsible for obtaining and maintaining the MSDSs;

2. How such sheets are to be maintained in the workplace (e.g., in notebooks in the work area(s) or in a computer with terminal access), and how employees can obtain access to them when they are in their work area during the work shift;

3. Procedures to follow when the MSDS is not received at the time of the first shipment;

4. For producers, procedures to update the MSDS when new and significant health information is found; and,

5. Description of alternatives to actual data sheets in the workplace, if used.

For employers using hazardous chemicals, the most important aspect of the written program in terms of MSDSs is to ensure that someone is responsible for obtaining and maintaining the MSDSs for every hazardous chemical in the workplace. The list of hazardous chemicals required to be maintained as part of the written program will serve as an inventory. As new chemicals are purchased, the list should be updated. Many companies have found it convenient to include on their purchase orders the name and address of the person designated in their company to receive MSDSs.

C. "Employee Information and Training"

Each employee who may be "exposed" to hazardous chemicals when working must be provided information and trained prior to initial assignment to work with a hazardous chemical, and whenever the hazard changes.

"Exposure" or "exposed" under the rule means that "an employee is subjected to a hazardous chemical in the course of employment through any route of entry (inhalation, ingestion, skin contact or absorption, etc.) and includes potential (e.g., accidental or possible) exposure." See Paragraph (h) of this section for specific requirements. Information and training may be done either by individual chemical, or by categories of hazards (such as flammability or carcinogenicity). If there are only a few chemicals in the workplace, then you may want to discuss each one individually. Where there are large numbers of chemicals, or the chemicals change frequently, you will probably want to train generally based on the hazard categories (e.g., flammable liquids, corrosive materials, carcinogens). Employees will have access to the substance-specific information on the labels and MSDSs.

Information and training is a critical part of the hazard communication program. Information regarding hazards and protective measures are provided to workers through written labels and material safety data sheets. However, through effective information and training, workers will learn to read and understand such information, determine how it can be obtained and used in their own workplaces, and understand the risks of exposure to the chemicals in their workplaces as well as the ways to protect themselves. A properly conducted training program will ensure comprehension and understanding. It is not sufficient to either just read material to the workers, or simply hand them material to read. You want to create a climate where workers feel free to ask questions. This will help you to ensure that the information is understood. You must always remember that the underlying purpose of the HCS is to reduce the incidence of chemical source illnesses and injuries. This will be accomplished by modifying behavior through the provision of hazard information and information about protective measures. If your program works, you and your workers will better understand the chemical hazards within the workplace. The procedures you establish regarding, for example, purchasing, storage, and handling of these chemicals will improve, and thereby reduce the risks posed to employees exposed to the chemical hazards involved. Furthermore, your workers' comprehension will also be increased, and proper work practices will be followed in your workplace.

If you are going to do the training yourself, you will have to understand the material and be prepared to motivate the workers to learn. This is not always an easy task, but the benefits are worth the effort. More information regarding appropriate training can be found in OSHA Publication No. 2254 which contains voluntary training guidelines prepared by OSHA's Training Institute. A copy of this document is available from OSHA's Publications Office at (202) 219-4667.

In reviewing your written program with regard to information and training, the following items need to be considered:

1. Designation of person(s) responsible for conducting training;

2. Format of the program to be used (audiovisuals, classroom instruction, etc.);

3. Elements of the training program (should be consistent with the elements in paragraph (h) of this section);

4. Procedure to train new employees at the time of their initial assignment to work with a hazardous chemical, and to train employees when a new hazard is introduced into the workplace.

The written program should provide enough details about the employer's plans in this area to assess whether or not a good faith effort is being made to train employees. OSHA does not expect that every worker will be able to recite all of the information about each chemical in the workplace. In general, the most important aspects of training under the HCS are to ensure that employees are aware that they are exposed to hazardous chemicals, that they know how to read and use labels and material safety data sheets, and that, as a consequence of learning this information, they are following the appropriate protective measures established by the employer. OSHA compliance officers will be talking to employees to determine if they have received training, if they know they are exposed to hazardous chemicals, and if they know where to obtain substance-specific information on labels and MSDSs.

The rule does not require employers to maintain records of employee training, but many employers choose to do so. This may help you monitor your own program to ensure that all employees are appropriately trained. If you already have a training program, you may simply have to supplement it with whatever additional information is required under the HCS. For example, construction employers that are already in compliance with the construction training standard (29 CFR 1926.21) will have little extra training to do.

An employer can provide employees information and training through whatever means are found appropriate and protective. Although there would always have to be some training on-site (such as informing employees of the location and availability of the written program and MSDSs), employee training may be satisfied in part by general training about the requirements of the HCS and about chemical hazards on the job which is provided by, for example, trade associations, unions, colleges, and professional schools. In addition, previous training, education and experience of a worker may relieve the employer of some of the burdens of informing and training that worker. Regardless of the method relied upon, however, the employer is always ultimately responsible for ensuring that employees are adequately trained. If the compliance officer finds that the training is deficient, the employer will be cited for the deficiency regardless of who actually provided the training on behalf of the employer.

D. "Other Requirements"

In addition to these specific items, compliance officers will also be asking the following questions in assessing the adequacy of the program: Does a list of the hazardous chemicals exist in each work area or at a central location?

Are methods the employer will use to inform employees of the hazards of non-routine tasks outlined?

Are employees informed of the hazards associated with chemicals contained in unlabeled pipes in their work areas?

On multi-employer worksites, has the employer provided other employers with information about labeling systems and precautionary measures where the other employers have employees exposed to the initial employer's chemicals?

Is the written program made available to employees and their designated representatives?

If your program adequately addresses the means of communicating information to employees in your workplace, and provides answers to the basic questions outlined above, it will be found to be in compliance with the rule.

5. "Checklist for Compliance"

The following checklist will help to ensure compliance with the rule:

Obtained a copy of the rule. ______________

Read and understood the requirements. ______________

Assigned responsibility for tasks. ______________

Prepared an inventory of chemicals. ______________

Ensured containers are labeled. ______________

Obtained MSDS for each chemical. ______________

Prepared written program. ______________

Made MSDSs available to workers. ______________

Conducted training of workers. ______________

Established procedures to maintain current program. ______________

Established procedures to evaluate effectiveness. ______________

6. "Further Assistance"

If you have a question regarding compliance with the HCS, you should contact your local OSHA Area Office for assistance. In addition, each OSHA Regional Office has a Hazard Communication Coordinator who can answer your questions. Free consultation services are also available to assist employers, and information regarding these services can be obtained through the Area and Regional offices as well.

The telephone number for the OSHA office closest to you should be listed in your local telephone directory. If you are not able to obtain this information, you may contact OSHA's Office of Information and Consumer Affairs at (202) 219-8151 for further assistance in identifying the appropriate contacts.

# Hazard Communication Training and Testing

## HazCom Training

Effective training is the most important part of a successful program. Each employee must completely understand his/her role for compliance. There are several ways to accomplish the necessary training. A classroom style or module-based training may be used. Since most of the information the employee will need to know is provided in the program itself, use it! Of course, you will need to add some important information such as the customized modifications that you've added to the program or job function specifics. Any information you can offer the employee to increase knowledge of the products and communication effort within the facility is beneficial. Don't short-step your program for time convenience.

Other ways to accomplish more awareness include adding training information for items such as material handling that apply to your facility. For instance, if your operation requires the use of "Acid X", explain the methods for transporting, transferring, spill cleanup, etc. Even if you have a material-handling program or similar that demonstrates such procedures, remember that certain safety concerns can never be overemphasized. Also, consider using a good hazard communication video in addition to the written training.

Decide the method of training that will best suit the needs of the company and its employees. Again, there are several factors to consider: company size, processes used, degree of hazard posed by chemicals used, etc. Because of the varying factors that make up the HazCom program and training, don't forget that it is the responsibility of the employer to determine training necessities; we can only offer suggestions.

## Checklist for Employee Training

Use the following checklist to provide employee training (the sample training module provides additional information concerning each subject):

- ❑ Employees shall understand that they have the right to access any of the following information:
  - Chemicals used in the workplace
  - Results of tests or analyses
  - Effects that chemicals or the environment may have on their health
  - MSDSs
- ❑ Employees are trained upon initial employment, annually, and when the employee is exposed to a new or altered chemical hazard.
- ❑ Training should include the following topics:
  - Methods used to detect the presence of hazardous chemicals
  - Physcial and health hazards of chemicals used
  - Measures for protection from exposure
  - "Right-to-Know"
  - Company's adopted Hazardous Material Identification System (HMIS)
  - How to read and understand an MSDS
  - Location of MSDSs

## Testing

Once an employee has completed the training module, a quiz should follow to determine the employee's knowledge of the program and company policy. It is not necessary to issue the employee a letter or numerical grade for the test. However, it is the employer's responsibility to review incorrect answers with the employee to ensure a complete understanding. It is also the employer's responsibility to determine (from testing) if the employee is capable of performing functions in a safe manner regarding hazardous materials. This may be accomplished by retesting or job reclassification. If unsure about an employee's ability to perform a job function safely, do not allow the employee to continue. Implement company policy that demonstrates the seriousness of your commitment for safety. A sample test follows the training module in this section.

## Training Information to Be Completed by Company

Complete the following information in the spaces provided to adapt the sample training module to your company:

- ❑ Locations of the HazCom Program Manuals.
- ❑ Vendor, position, or department that should provide the labels used at your facility.

> The sample Training Module (pages 57-83) may be excerpted in its entirety to serve as a template for your company's training program. You may also obtain an electronic version of this template by calling Government Institutes at 301-921-2355.

# Hazard Communication Training Module

## Contents

## Introduction

The purpose of this training module is to offer employees required to handle hazardous materials adequate training for the proper handling, labeling, and storage of hazardous materials. This training module also fills OSHA requirements for training each HazMat employee of their "Right-to-Know" (Hazard Communication: 29 CFR Part 1910.1200).

This module will begin by offering some basic definitions and acronyms, followed by a section that includes the module's application. General Awareness, Function Specific, and Safety requirements have also been included to fill the regulation's requirements for each.

It is the responsibility of each HazMat employee to use the provided information wisely, including the Workplace Chemical List, MSDSs, and the Hazardous Materials Identification System as explained in the Hazard Communication Program.

## Definitions and Acronyms

- *Carrier* means a person engaged in the transportation of passengers or property by:
  - Land or water, as a common, contract, or private carrier, or
  - Civil aircraft.
- *C* means Celsius or Centigrade.
- *Carcinogen* means a substance or agent capable or causing cancer in mammals, including humans.
- *CC* means closed cup.
- *CFR* means Code of Federal Regulations.
- *Class* means hazard class. See *Hazard Class*.
- *Combustible Liquid* means any liquid that does not meet the definition of any other hazard class specified in 49 CFR Part 173.120 and has a flash point above 60.5°C (141°F) and below 93°C (200°F). A flammable liquid that does not meet the definition of any other hazard class may be re-classed as a combustible liquid. This provision does not apply to transportation by vessel or aircraft, except where other means of transportation are impracticable.
- *Consumer Commodity* means a material that is packaged and distributed in a form intended or suitable for sale through retail sales agencies for consumption by individuals for purposes of personal care or household use. This term also includes drugs and medicines.
- *Corrosive Material* means a liquid or solid that causes visible destruction or irreversible alterations in human skin tissue at the site of contact, or a liquid that has a severe corrosion rate on steel or aluminum in accordance with the criteria in 49 CFR 173.136.

- *DOD* means the U.S. Department of Defense.
- *EPA* means the U.S. Environmental Protection Agency.
- *F* means degrees Fahrenheit.
- *Flammable Gas* means any material which is a gas at 20°C (68°F) or less and 101.3 kPa (14.7 psi) of pressure which:
  - is ignitable at 101.3 kPa (14.7 psi) when in a mixture of 13% or less by volume with air; or
  - has a flammable range at 101.3 kPa (14.7 psi) with air of at least 12% regardless of the lower limit.
- *Flammable Liquid* means a liquid having a flash point of not more than 60.5°C (141°F) or any material in a liquid phase with a flash point at or above 37.8°C (100°F) that is intentionally heated and offered for transportation or transported in bulk packaging.
- *Flammable Solid* means solids which may explode when wetted, are self-reactive materials, or are readily combustible solids which may cause fire through friction, such as matches.
- *Flash Point* means the minimum temperature at which a liquid gives off vapor within a test vessel in sufficient concentration to form an ignitable mixture with air near the surface of the liquid.
- *Hazard Communication* means the employer's attempt to communicate the hazards present to the everyday duties of the Hazmat employee. This is done so by means of Material Safety Data Sheets (MSDSs), the employer's hazardous materials identification system (HMIS), labels, etc.
- *Hazard Class* means the category of hazard assigned to a hazardous material under the definitional criteria of 49 CFR 173. A material may meet the defining criteria of more than one hazard class but is assigned to only one hazard class.
- *Hazardous Material* means a substance or material which has been determined by the Secretary of Transportation to be capable of posing an unreasonable risk to health, safety, and property when transported in commerce, and which has been so designated. The term includes hazardous substances, marine pollutants, and elevated temperature materials. A hazardous material may also be defined as any material that meets any of the hazard class definitions in 49 CFR 173.
- *Hazmat Employee* means an employee of a Hazmat employer who during the course of employment loads, unloads, or handles hazardous materials; tests, reconditions, repairs, modifies, marks, or otherwise represents containers, drums, or packagings as qualified for the use of transportation of hazardous materials; prepares hazardous materials for transportation; is responsible for safety in transporting hazardous materials; or operates a vehicle used to transport hazardous materials.
- *HMIS* means hazardous materials identification system.
- *Inert Material* means a material that is not combustible for purposes of spill cleaning and/or absorption (e.g. peat moss, sand, etc.)

- *MSDS* means Material Safety Data Sheet.
- *NFPA* means National Fire Protection Agency.
- *Non-flammable Gas* means any material which exerts in the packaging an absolute pressure of 280 kPa (41 psia) or greater at 20°C (68°F), and does not meet the definition of Division 2.1 or 2.3.
- *N.O.S.* means not otherwise specified.
- *ORM* means other regulated material.
- *OSHA* means Occupational Safety and Health Administration.
- *Psi* means pounds per square inch.
- *Psia* means pounds per square inch absolute.
- *UN* means United Nations.

## Application

- Training - Federal regulations require specific levels of training to be provided to HazMat employees. These levels of training are General Awareness, Function Specific, and Safety.
- Hazard Communication - Hazardous materials communication requirements are fulfilled by the use of shipping papers, markings, labels, placards, emergency response information, incident reporting, and in some cases, the packing group (DOT/IATA). The Hazard Communication Program has been developed to inform personnel of any hazards, dangers, or risks that may occur while handling or working with hazardous or toxic materials.
- Transportation – This training program does not qualify individuals to ship or offer for shipment hazardous materials. Only those personnel who have been trained and certified to ship hazardous materials should do so.
- Testing – All HazMat employees must be tested on the training requirements of this module. Upon completion of testing, the trainer will review the test to cover incorrectly answered questions and offer time to answer any other questions.

# Hazard Communication

## I. Purpose and Application

### A. Scope

- All personnel must receive training on hazardous chemicals in their assigned work area when they are first employed. Persons who have not received this training will not be permitted to work in areas where hazardous chemicals or substances exist.
- All personnel must receive re-training within 12 months of initial training and at 12-month intervals thereafter. This training must include any new hazard information, if applicable.
- Local management will administer this training module. Upon review of the module, each student must complete the quiz and return to his/her supervisor for correction and discussion. The student and supervisor must sign the quiz indicating the student understands the program and any missed questions. The most recent quiz will be maintained on local file for one year.

### B. Hazard Communication (Title 29 CFR 1910.1200)

Hazardous chemicals, hazardous materials, or hazardous substances are used for many purposes in homes, businesses, and industry. Often people who use such materials are unaware of specific hazards or hazardous properties, and what should be done to protect themselves from illness or injury chemicals may cause when used improperly.

Approximately 25 million workers (about one in four in the nation's work force) are exposed to one or more potential chemical hazards. There are an estimated 575,000 chemical products on the market with hundreds of new ones being introduced annually.

Chemical exposure may cause or contribute to many serious health effects such as heart ailments, kidney and lung damage, sterility, cancer, burns, and rashes. Some chemicals may also be safety hazards and have the potential to cause fires, explosions, or other serious accidents.

Because of the seriousness of these potential safety and health hazards, OSHA announced its Federal Hazard Communication Standard (Title 29 CFR 1910.1200) in 1983. The standard initially applied only to chemical manufacturers, importers, distributors, and other manufacturing industries. However, on August 24, 1987, OSHA revised the Hazard Communication Standard by extending the standard to include all companies employing personnel who are, or may be, exposed to potentially hazardous chemicals, materials, or substances.

The purpose of the Hazard Communication Standard is to establish uniform requirements to ensure the employee is informed of all potentially hazardous chemicals used in the workplace.

## II. Basic Concepts of Safety

### A. Safety

*Safety* is defined as "the state of being safe" and is commonly thought of as "freedom from danger or harm."

A large part of the Hazard Communication Standard is to provide reliable information about chemical and material hazards so chemical handlers have a better understanding of the potential risks, make better decisions, and thus, exercise better precautions regarding overall safety.

By exercising good safety habits, the probability of becoming injured or contracting an illness from a particular chemical or substance in the work place is extremely small.

For many years, most large chemical manufacturing companies have had safety records that are far better than overall industry averages. These safety records have been achieved in spite of the fact that chemical industry workers have a much greater exposure to hazardous materials than workers in most other industries. This is largely the result of a better understanding of the hazardous properties of the materials; therefore, the workers are often better motivated to practice safer work habits.

### B. Hazards

The dictionary defines a *hazard* as

- ❑ A chance happening
- ❑ A chance of being harmed or injured
- ❑ A possible source of danger

A hazard may also be defined as any substance, situation, or condition capable of causing harm to human health, property, or system functioning. Notice particularly that these definitions do not state that a hazard will cause harm, but merely states that the capability, possibility, or chance of harm exists. Thus, by exercising safe work habits and taking the proper precautions, the chance of a hazardous substance causing harm is unlikely.

### C. Risk

The dictionary defines *risk* as

- ❑ The possibility of suffering harm or loss
- ❑ A factor, course, or element involving uncertain danger
- ❑ To expose to a chance of loss or harm

A risk may also be defined as a measure of the probability and severity of harm to human health, property, or system functioning. In other words, "risk" includes a sense of how harm is likely to occur. Notice also, that these definitions do not state that a risk will cause harm or loss, but merely states that the possibility or chance of harm or loss exists.

# III. Material Safety Data Sheets

A Material Safety Data Sheet (MSDS) is a document that describes the physical and chemical properties of products, their physical and health hazards, and precautions for safe handling and use.

The manufacturers and distributors of chemical products furnish MSDSs to ***The Company***. An MSDS must be maintained for each hazardous chemical used. All departments should coordinate activities to ensure that all incoming initial shipments of hazardous substances arrive with an MSDS. While MSDSs are not required to be physically attached to the shipment, they must accompany or precede the shipment. In cases where a department receives repetitive shipments from the same supplier, the MSDS may already be on file. Although subsequent shipments of the same item may be accompanied with an MSDS, An MSDS is only required with the initial shipment. However, all subsequent MSDSs should be reviewed for accuracy and a comparison of old and new MSDSs should be conducted. This is to identify those situations where a "new" hazard associated with an existing chemical has been identified, or a new ingredient is included in a currently used product. Chemical manufacturers, importers, and distributors are required to add significant information concerning a product to an MSDS within 90 days.

**All MSDSs, initially or subsequently, must be copied and forwarded to the Program Coordinator when received. Once the MSDS has been received and reviewed, it should be assigned a number for reference and returned. This numbered MSDS should be inserted into the local MSDS manual, as appropriate, replacing the previous sheet.**

The HazCom Standard does not describe the format to be used in completing the MSDS. However, the standard does specify the information the MSDS must contain (Title 29 CFR 1910.1200(g)). OSHA Form 174 meets the HazCom Standard's requirements and is recommended (See Figure 3-1).

The HazCom Program manuals, which contain all MSDS files, are readily accessible to all employees during each work shift and are located at the following locations:

- **Master MSDS file -** ______________________________
- ______________________________
- ______________________________
- ______________________________

| Material Safety Data Sheet | U.S. Department of Labor |
|---|---|
| May be used to comply with<br>OSHA's Hazard Communication Standard,<br>29 CFR 1910.1200. Standard must be<br>consulted for specific requirements. | Occupational Safety and Health<br>(Non-Mandatory Form)<br>Form Approved<br>OMB No. 1218-0072 |

**A**

| IDENTITY (As Used on Label and List) | Note: Blank spaces are not permitted. If any is not applicable, or no information is available, the space must be marked to indicate that. |
|---|---|

**Section I**

**B**

| Manufacturer's Name | Emergency Telephone Number |
|---|---|
| Address (Number, Street, City, State, and ZIP Code) | Telephone Number for Information |
| | Date Prepared |
| | Signature of Preparer |

**C**

**Section II --- Hazardous Ingredients/Identity Information**

| Hazardous Components (Specific Chemical Identity: Common Name(s)) | OSHA PEL | ACGIH TLV | Other Limits Recommended | % (optional) |
|---|---|---|---|---|
| | | | | |

**D**

**Section III --- Physical/Chemical Characteristics**

| | | | |
|---|---|---|---|
| Boiling Point | | Specific Gravity (H2O = 1) | |
| Vapor Pressure (mm Hg.) | | Melting Point | |
| Vapor Density (AIR = 1) | | Evaporation Rate (Butyl Acetate = 1) | |
| Solubility in Water | | | |
| Appearance and Odor | | | |

**E**

**Section IV --- Fire and Explosion Hazard Data**

**F**

| | | | |
|---|---|---|---|
| Flash Point (Method Used) | Flammable Limits **G** | LEL **H** | UEL |
| Extinguishing Media | | | |
| Special Fire Fighting Procedures | | | |
| Unusual Fire and Explosion Hazards | | | |

**I**

(Reproduce locally) OSHA 174, Sept. 1985

**Figure 3-1: Sample MSDS (Front)**

J K L M N O P Q R S T U V W X

## Section V --- Reactivity Data

| Stability | Unstable | | Conditions to Avoid |
|---|---|---|---|
| | Stable | | |
| Incompatibility (Materials to Avoid) | | | |
| Hazardous Decomposition or Byproducts | | | |
| Hazardous Polymerization | May Occur | | Conditions to Avoid |
| | Will Not Occur | | |

## Section VI --- Health Hazard Data

| Route(s) of Entry: | Inhalation? | Skin? | Ingestion? |
|---|---|---|---|
| Health Hazards (Acute and Chronic) | | | |
| Carcinogenicity: | NTP? | IARC Monographs? | OSHA Regulated? |
| Signs and Symptoms of Exposure | | | |
| Medical Conditions Generally Aggrivated by Exposure | | | |
| Emergency and First Aid Procedures | | | |

## Section VII --- Precautions for Safe Handling and Use

| Steps to Be Taken in Case Material Is Released or Spilled |
|---|
| Waste Disposal Method |
| Precautions to Be Taken in Handling and Storing |
| Other Precautions |

## Section VIII --- Control Measures

| Respiratory Protection (Specify Type) | | |
|---|---|---|
| Ventilation | Local Exhaust | Special |
| | Mechanical (General) | Other |
| Protective Gloves | | Eye Protection |
| Other Protective Clothing or Equipment | | |
| Work/Hygenic Practices | | |

**Figure 3-1 (cont.): Sample MSDS (Back)**

# IV. MSDS (OSHA Form 174) Explanation

A. Section I – Chemical Identification – The identity of the material must indicate specific chemical name, common name, and trade name, if applicable. The specific chemical name is one that will permit any qualified chemist to identify the structure of the compound(s) present to enable further searching of literature for desired information.

B. Name and address of the chemical manufacturer. Phone number to be used for information or an emergency.

C. Section II – Hazardous Ingredients/Identity Information – Lists content of the substance that may cause harm. The chemical name(s) of the substance, as well as the common name of the substance must be indicated. It also lists the concentration of the chemicals to which you can safely be exposed to within an eight-hour period (PEL – permissible exposure limit, or TLV – threshold limit value).

D. Section III – Physical/Chemical Characteristics – Describes the substance's appearance, odor, and other chemical characteristics. These are things that can affect the degree of hazard workers face in different work situations.

- Boiling Point/Melting Point – to help prevent a potentially dangerous change in state, as from a liquid to a breathable gas.
- Vapor Pressure, Density, and Evaporation Rate – important for flammable or toxic gases and vapors.
- Solubility in Water and Specific Gravity – identifies whether a chemical will dissolve in water, sink, or float.

E. Normal Appearance and Odor – Helps you recognize anything different and possibly dangerous.

F. Section IV – Flash Point – Indicates the minimum temperature at which a liquid gives off a vapor in sufficient concentration to ignite.

G. Flammable Limits – Indicates the concentration of the substance, in the form of a gas or a vapor that is required for it to ignite.

H. LEL and UEL – Lowest and highest (respectively) concentration (percentage of substance in the air) that will produce a flash of fire when an ignition source is present. At higher concentrations, the mixture is too "rich" to burn.

I. This section describes methods of firefighting including the proper extinguishing media (i.e. ABC, $CO_2$, foam, etc.). This section also describes unusual fire and explosion hazards.

J. Section V – Reactivity Data – Indicates whether the substance is stable or unstable and conditions to avoid, such as direct sunlight, or any condition which may cause a dangerous reaction.

K. Incompatibility – Lists materials, chemicals, and other substances to avoid which may cause the chemical to burn, explode, or release dangerous gases.

L. Section VI – Health Hazard Data – This is probably the most important section of the MSDS. It describes how a substance can enter your body, for instance:

- Inhalation (respired)
- Through the skin (absorbed)
- Swallowing (consumed)

M. Health Hazards (Acute and Chronic) – Lists the specific possible health hazards which could result from exposure, both acute and chronic.

Some effects like minor skin burns, are acute (they show up immediately after exposure).

Others, like liver damage, are chronic (they are often the result of exposure long ago or repeated small exposures over a long period of time).

N. If the substance or chemical is considered or believed to be a carcinogen, it will be indicated here.

O. This section lists the signs and symptoms of overexposure such as eye irritation, nausea, dizziness, skin rashes, headache, and burns.

P. If exposure to the substance could aggravate an existing medical condition, this information will be indicated here.

Q. This section describes emergency and first aid procedures to be followed in case of overexposure until medical help arrives. If these instructions are unclear, seek medical assistance immediately.

R. Section VII – Precautions for Safe Handling and Use – Explains procedures for spills, leaks, or any release; how to handle and store the substance; how to safely dispose of the substance; and other precautions.

S. This section describes the proper procedures to be followed in the case of spills or leaks, and the equipment needed.

*Note:* *Not all MSDSs will indicate the personal protective equipment recommended when cleaning up a spill. Always refer to section VIII – Control Measures, for specific types of recommended personal protective equipment to be worn prior to cleaning up a spill.*

Always notify your supervisor when a chemical spill or leak occurs, no matter what chemical, no matter how small.

T. This section lists the proper disposal procedures.

U. This section describes the proper handling and storage procedures, as well as other precautions to be followed to protect you and others (e.g. grounding containers during transfer of flammable materials to prevent static electricity as an ignition source).

V. Section VIII – Control Measures – Lists control measures for preventing or reducing the chance of harmful exposure, including personal protective equipment to be worn, work and hygiene practices, and ventilation requirements.

W. This section includes proper respiratory equipment and ventilation information.

X. This section describes gloves, eye protection, and other personal protective clothing recommended or required.

This section lists work or hygiene practices which should be exercised when handling this substance or chemical such as:

- Taking a shower
- Washing work clothes
- Destroying soiled clothing
- Washing hands

## You control the risks!!!

The Material Safety Data Sheet has been created and is required by law as a part of the "Right-To-Know" Program.

***The Company*** goes to great trouble to keep them on file for you. MSDSs are your personal safety guide to the hazardous substances/materials you work with.

## Read them!!!

Check the MSDS prior to using any hazardous chemical you are unfamiliar with. Follow the procedures and use recommended equipment.

You will not find the exact same information on every MSDS. However, you will be informed of all information known about the chemical, its hazards, and the precautions you can take to avoid injury and illness when handling a hazardous substance.

## V. Workplace Chemical List

A. General – The Workplace Chemical List identifies stock materials that are or contain ingredients classified by OSHA and/or the DOT as potentially hazardous to personnel. The Hazard Communication Standard requires that a list of hazardous chemicals be included as part of the company's Hazard Communication Program. The list will serve as an inventory for which an MSDS is maintained on file. Supplemental information contained on the Workplace Chemical List is intended only for the convenience or quick reference of the employee.

Workplace Chemical Lists are located behind the Hazard Communication Program. The Program Coordinator prepares the list with information compiled from the respective MSDS supplied by the manufacturer of the hazardous material.

B. Sample – Figure 3-2 shows a Sample Workplace Chemical List, Section C provides a detailed explanation of the rows on the list. This list will give information that will assist in the safe handling and proper labeling of materials. The lists in the workplace are preceded by an index identifying the letters assigned to the specified rows (these may be found in Tables 3-1 through 3-5).

| 1 | 2 | 3 | 4 | 5 | 6 | 7 | 8 | 9 | 10 |
|---|---|---|---|---|---|---|---|---|---|
| MSDS# | Description | MSDS Date | Health Hazard | PPE | Flashpoint | First Aid | Spill | Disposal | Location |
| 1000 | Acetone, ABC Chemical Corp. | August 1998 | Health – 1<br>Fire – 3<br>Reactivity Specific – 0 | B | 5°F | Eyes – A<br>Skin – A<br>Inhale – A<br>Ingest – A | B | A | Paint Shop |
| 1120 | Lacquer Thinner, Best Paint Company | December 1998 | Health – 2<br>Fire – 3<br>Reactivity Specific – 0 | H | 21°F | Eyes – A<br>Skin – A<br>Inhale – A<br>Ingest – A | A | A | Paint Shop |
| 1230 | Methyl Ethyl Ketone | April 1998 | Health – 3<br>Fire – 3<br>Reactivity Specific – 0 | I | 23°F | Eyes – A<br>Skin – A<br>Inhale – A<br>Ingest – A | B | A | All |

## Figure 3-2: Sample Workplace Chemical List

C. Column Explanations

1. MSDS Number – Identifies the product listed in the second column with an MSDS number. This will serve as the product's permanent reference number.

2. Description – The name of the chemical and, in some cases, a brief description of the product.

3. MSDS Date – The date the MSDS was produced.

4. Health Hazard Rating – The health hazard rating (NFPA rating system) for the product listed in the second column. This number rating is defined in Table 3-1.

   Fire Hazard Rating – The flammable rating (NFPA rating system) for the product listed in the second column. This number rating is defined in Table 3-1.

   Reactivity Hazard Rating – The reactivity hazard rating (NFPA rating system) for the product listed in the second row. This number rating is defined in Table 3-1.

5. PPE – A letter reference as to the personal protective equipment recommended for working with the product listed in the second column. See Table 3-2 for a definition of the referenced letters.

6. Flashpoint – Refers to the temperature, usually in degrees *F*, at which the total product listed in the second column, is capable of producing enough flammable vapors to ignite. Products having a flash point below 100°F are classified as flammable and should be handled with special precautions to prevent combustion.

7. First Aid – Gives a letter reference as to the measures that should be taken when an accidental contamination of the eyes, skin, lungs, or intestines has occurred. See Tables 3-3-1 through 3-3-4 for an index of the referenced letters.

8. Spill – Gives a letter reference as to the measures to be taken when an accidental spill occurs. See Table 3-4 for an index of the referenced letters.

9. Disposal – Gives a letter reference as to the measures to be taken for the proper disposal of the material listed in the second column. See Table 3-5 for an index of the referenced letters.

10. Location – Indicates the location where the product may be found within the facility.

D. Index for Workplace Chemical List - The following tables explain and/or describe the letters assigned to the different rows of the Workplace Chemical List. Included are indices for personal protective equipment, first aid, spill cleanup, and disposal measures. These tables may be found in the front of the Workplace Chemical List at each listed location. There is also a Personal Protection Index posted in various locations throughout all facilities where hazardous materials are present.

## Table 3-1: Hazard Index

**Health Hazard (Blue)**

4 - Deadly; short exposure may cause death or serious injury

3 - Extremely dangerous; short exposure could cause serious, temporary, or residual injury

2 - Hazardous; intense or continued exposure could cause temporary incapacitation or possible residual injury

1 - Slightly hazardous; on exposure, could cause irritation, but only minor residual injury

0 - No hazard

**Reactivity Hazard (Yellow)**

4 - Extremely reactive, may detonate

3 - May detonate with shock and heat

2 - Chemical change may be violent

1 - Unstable if heated

0 - Stable

**Fire Hazard (Red)**

4 - Extremely flammable, flash point below 0°F

3 - Highly flammable, flash point 1°F-100°F

2 - Moderately flammable, flash point 101°F-200°F

1 - Slightly flammable, flash point above 201°F

0 - Not flammable

**Basic Hazard Index**

4 - Severe Hazard

3 - Serious Hazard

2 - Moderate hazard

1 - Slight Hazard

0 - Minimal Hazard

## Table 3-2: Personal Protection Index

A. Safety Glasses
B. Safety Glasses, Chemical Gloves
C. Safety Glasses, Chemical Gloves, Synthetic Apron
D. Face Shield, Chemical Gloves, Synthetic Apron
E. Safety Glasses, Chemical Gloves, Dust Respirator
F. Safety Glasses, Chemical Gloves, Synthetic Apron, Dust Respirator
G. Safety Glasses, Chemical Gloves, Vapor Respirator
H. Splash Goggles, Chemical Gloves, Synthetic Apron, Vapor Respirator
I. Safety Glasses, Chemical Gloves, Dust and Vapor Respirator
J. Splash Goggles, Chemical Gloves, Synthetic Apron, Dust and Vapor Respirator
K. Airline Hood or Mask, Chemical Gloves, Full Suit, Rubber Boots
X. Ask your supervisor for special handling instructions

First Aid – Tables 3-3-1 through 3-3-4 define the emergency procedures recommended for the treatment of an individual who has experienced an injury or exposure to a particular hazardous material. These first aid measures have been taken from the MSDSs and are considered as interim procedures only to be administered between exposure and treatment by licensed/trained medical personnel. These measures in no way are intended to replace or supersede professional treatment in the case of injury or exposure. A licensed physician should examine an injured employee as soon as possible.

## Table 3-3-1: First Aid – Eyes

A. Flush with water for 15 minutes. Get medical attention.
B. Flush with plenty of water.
C. Flush with lukewarm water for 15 minutes. Get medical attention.
D. Flush with water until irritation subsides.
E. Flush with warm water. Apply gauze patch. Get medical attention.
F. Flush with cold water. Get medical attention.
G. Flush with water for 15 minutes.
H. No first aid should be needed.
X. Reference respective MSDS number for specific first aid information.

## Table 3-3-2: First Aid – Skin

A. Wash with soap and water. Contact physician if irritation persists.
B. Wash with soap and water. Apply skin cream that contains lanolin.
C. Flush with water for 15 minutes.
D. Flush with water. Contact physician if irritation persists.
E. Soak in warm soapy water.
F. Wipe off product and flush with water.
G. Wash with soap and water. Get medical attention.
H. Wipe off contamination. Wash with soap and water.
I. Flush with lukewarm water.
J. No first aid should be needed.
X. Reference respective MSDS number for specific first aid information.

## Table 3-3-3: First Aid – Inhalation

A. Remove to fresh air. If breathing has stopped, give mouth-to-mouth. Get medical attention.
B. Remove to fresh air. Keep warm and quiet. Get medical attention.
C. No first aid treatment should be necessary.
X. Reference respective MSDS number for specific first-aid information.

## Table 3-3-4: First Aid – Ingestion

A. Do not induce vomiting. Keep warm and quiet. Get medical attention.
B. Rinse mouth. Give plenty of water followed by milk, egg white, or gruel. Do not induce vomiting. Get medical attention.
C. Contact poison control center or physician immediately.
D. Give plenty of water. Get medical attention.
E. Give two (2) glasses of water. Induce vomiting. Get medical attention.
F. Do not induce vomiting. Give plenty of water. Get medical attention.
G. Give milk, water, or egg whites. Induce vomiting. Get medical attention.
H. Induce vomiting by giving a tablespoon of salt or mustard in a glass of warm water. Get medical attention.
I. Induce vomiting. Get medical attention.
J. Drink Milk of Magnesia, aluminum hydroxide gel, or lime water, followed by several glasses of water. Call physician. Do not induce vomiting.
K. Do not induce vomiting. Drink two (2) glasses of milk or water. Call physician or poison control center.
L. Rinse mouth with water several times.
M. No first aid should be needed.
X. Reference respective MSDS number for specific first aid information.

Spills – Table 3-3-4 describes the measures that should be taken to properly contain and clean up any material spilled.

## Table 3-4: Spill Containment and Cleanup

A. Absorb on inert material.
B. Eliminate ignition sources. Ventilate area. Absorb on inert material.
C. Absorb on inert material. Flush area with water.
D. Dike spill. Absorb on inert material. Sprinkle sodium thiosulfate on residue. Remove residue. Clean area with water.
E. Vacuum, scoop, or sweep discharged material.
F. Dilute by adding large volumes of water.
G. Neutralize with lime or soda ash. Absorb on inert material.
H. No spill or clean up measures should be necessary.
X. Reference respective MSDS number for specific spill containment and clean-up procedures.

Disposal – Table 3-3-5 defines procedures for waste disposal of the materials listed.

## Table 3-5: Waste Disposal Procedures

| | |
|---|---|
| A. | Consult local, state, and federal regulations. |
| B. | Do not incinerate container. Consult local, state, and federal regulations. |
| C. | Flush to sewer. |
| D. | There should be no residue for disposal. |
| E. | Dispose in trash. |

*Note:* *The EPA levies extremely high civil penalties for the improper disposal of hazardous/toxic substances. Consult local waste treatment control and disposal center and ensure compliance with all local, state, and federal regulations.*

## VI. Hazardous Materials Labeling

Federal regulations require all containers used in the storage and/or transportation of hazardous and toxic chemicals to be labeled, tagged, or marked with the following information:

- Identity of the hazardous chemical(s);
- Appropriate hazard warning; and,
- The name and address of the manufacturer.

In some cases, this regulation is satisfied by the warning information that is printed on the product label. However, there are some products that do not have this information. In such cases, the department manager/supervisor will be responsible for identifying and labeling such products.

The department manager/supervisor will be responsible for ensuring that all portable containers are labeled. Portable containers are defined as pressure sprayers, spray bottles, squeeze bottles, hydraulic servicing units, or any container used in the transfer or temporary storage of products which contain hazardous ingredients as defined in Title 29 CFR 1910.1200.

The Hazardous Materials Identification System (HMIS) is a standard system that communicates the hazards to the employee. The Hazardous Materials Identification Guide (HMIG), as it is referred to, communicates necessary hazard information by use of a common numerical system for Health (blue), Flammability (red), Reactivity (yellow), and Personal Protective Equipment required or specific hazard information (white).

In order to properly complete the HMIG label (see Figure 3-3), it is necessary to reference the Workplace Chemical List. In the list, find the product that is to be labeled. Indicate the name of the product on the top section of the label (if not legible on container). The Health, Flammability, and Reactivity section is rated by using the numerical ratings indicated in the Hazard Rating row. To indicate the recommended Personal Protective Equipment, use the letter that is given in the PPE row.

For an explanation of the Hazard Index, refer to Table 3-1. The explanation for the Personal Protection Index may be found in Table 3-2.

## VII. Health Hazard Characterization

Materials that have one or more of the following characteristic properties are covered by the Hazard Communication Standard:

| | | |
|---|---|---|
| irritations | cutaneous hazards | toxic agents |
| highly toxic agents | corrosives | eye hazards |
| blood acting agents | sensitizers | carcinogens |
| hematopietic system agents | hepatoxins | neurotoxins |
| reproductive toxins | nephrotoxins | |

agents that damage the lungs, skin, or mucous membranes

Health hazards are divided between acute health hazards and chronic health hazards. Some hazards have characteristics of both, but generally they are one or the other.

A. Acute Hazards – The effects of acute hazards are manifested soon after a single, brief exposure. Many acute effects disappear after a time and generally are not permanent. However, some may show permanent effects, and therefore can be considered both acute and chronic.

   Acute, when used to describe a hazard, does not imply severity. For the purpose of this program, the term acute refers to a short-term effect.

## Figure 3-3: Hazardous Materials Identification Guide

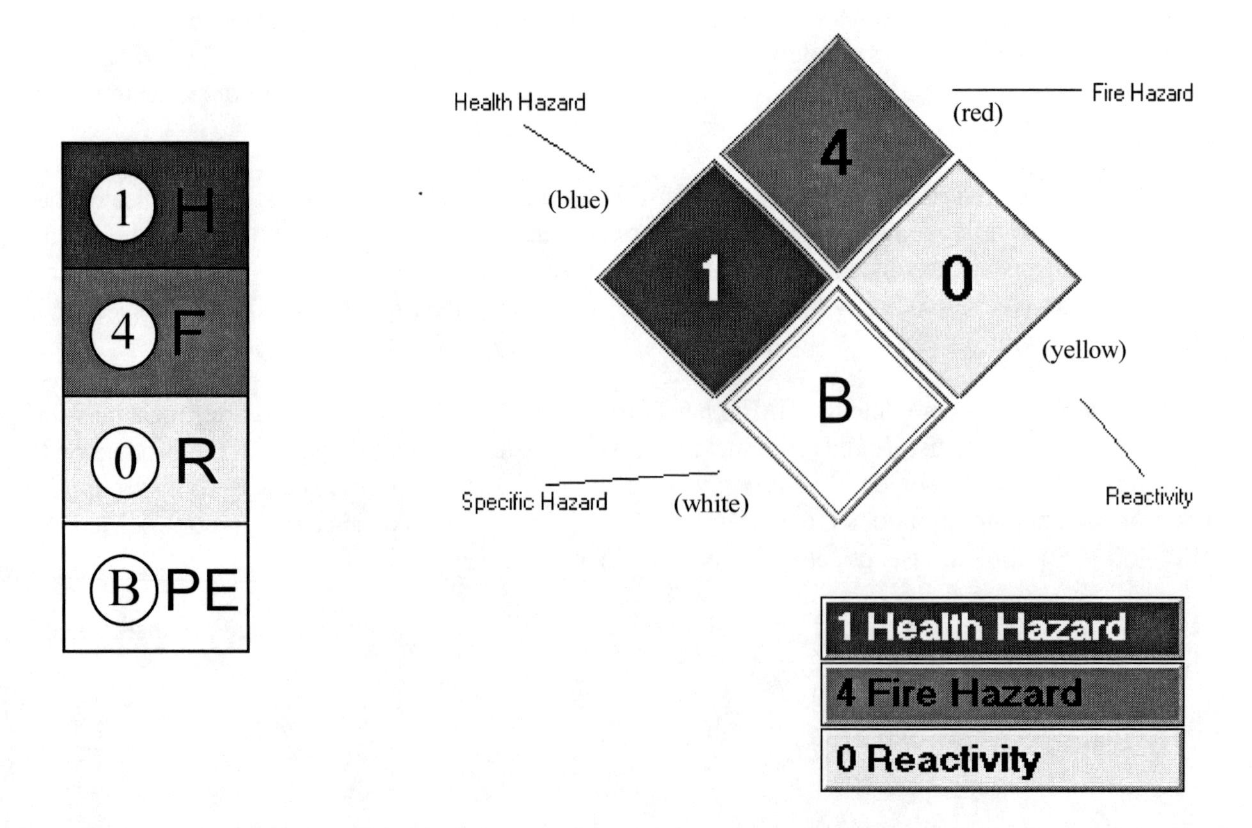

1. Irritants – An irritant is defined as a chemical capable of causing reversible inflammation at the site of contact by chemical action. Examples include nitric oxide, sodium hypochlorite, stannic chloride, and ethyl alcohol.

2. Cutaneous Hazards – A cutaneous hazard is a material that affects the dermal layer of the body, such as defatting of the skin, and causing rashes or skin irritation. Examples include acetone and chlorinated compounds.

3. Toxic Agents – Toxic agents are those defined as substances which can cause acute injury to the human body, or which are suspected of being able to cause diseases or injury under certain conditions.

4. Corrosive Materials – A corrosive material is a chemical causing visible destruction of, or irreversible alterations in, living tissue at the site of contact by chemical reaction. Examples include caustic soda, sulfuric acid, hydrofluoric acid, phenol, and boron trifluoride.

5. Eye Hazards – An eye hazard is a material that affects the eye or visual capacity, for example, by causing conjunctivitis or corneal damage. Examples include organic solvents, acids, and alkalis.

6. Agents that Act on the Blood or Hemalopoietic System – This type of agent is a substance that decreases the hemoglobin function and deprives the body tissue of oxygen. Cyanosis and loss of consciousness are typical symptoms. Examples include carbon monoxide, cyanides, metal carbonyls, nitrobenzene, hydroquinone, aniline, and arsine.

B. Chronic Hazards – Chronic hazards have a long-term affect, essentially permanent. Their affects may be slow to develop, and often result from repeated or continuous exposure over a long period of time.

1. Sensitizers – A sensitizer is a chemical that causes a substantial portion of exposed people or animals to develop an allergic reaction in normal tissue after repeated exposure. Examples include hydroquinone, bromine, platinum compounds, isocyanates, and ozone.

2. Carcinogens – A substance or agent capable of causing or producing cancer in mammals, including humans. Examples include asbestos, benzene, beryllium, lead chromate, formaldehyde, vinyl chloride, trichloroethylene, and carbon tetrachloride.

3. Reproductive Toxins (teratogens) – Substances that can cause birth defects or sterility.

4. Hepatoxins – A hepatoxin is a chemical that can cause liver damage, such as enlargement or jaundice. Examples include carbon tetrachloride, nitrosamines, vinyl chloride, chlorobenzene, trichloroethylene, chloroform, and ethyl alcohol.

5. Nephrotoxins – A nephrotoxin is a chemical that can cause kidney damage such as edema or proteinuria. Examples include halogenated hydrocarbons, uranium, vinyl chloride, trichloroethylene, and ethyl alcohol.

6. Neurotoxins – A neurotoxin is a chemical that causes primary toxic affects on the central nervous system, such as narcosis, behavioral changes, or a decrease in motor functions. Examples include mercury, carbon disulfide, ethyl alcohol, acetylene, manganese, thallium, and tetraethyl lead.

7. Agents that Damage the Lungs – These agents irritate pulmonary tissue, resulting in cough, tightness in the chest, and shortness of breath. Examples include silica, asbestos, cotton fibers, coal dust, and toluene diisocyanate.

## VIII. "Right-to-Know" Poster

The "Right-to-Know" poster must be displayed at each location where products are used that are considered hazardous by OSHA and state regulations. At each of these locations, the posters should be displayed on bulletin boards, near main employee entrances, and in employee break areas. These posters should not be removed or defaced at any time. A sample "Right-to-Know" poster (Figure 3-4) has been included on the following page.

# IT'S YOUR RIGHT-TO-KNOW

The Occupational Safety and Health Administration (OSHA) developed the Hazard Communication Standard (29 CFR Part 1910.1200) to provide information to employees about the chemicals they work with. Employees have both a NEED and a RIGHT to know the identities and hazards of the chemicals to which they are exposed.

## CHEMICAL LIST

Employers are required to maintain a complete list of chemicals present in the workplace. A copy of the complete chemical listing must be included with the written hazard communication program.

The identity of the chemical on this list must be the same as that on the chemical label and MSDS.

THE CHEMICAL LIST IS AVAILABLE TO EMPLOYEES UPON REQUEST

## MATERIAL SAFETY DATA SHEETS

The Material Safety Data Sheet (MSDS) is a written document that conveys detailed information about the hazards associated with a specific chemical. An MSDS has been collected for each hazardous chemical used in the workplace. THE MSDSs ARE IMMEDIATELY AVAILABLE FOR EMPLOYEE USE AT ALL TIMES. The identity of the chemical on the MSDS is the same as the product label. EACH MSDS INCLUDES THE FOLLOWING:

- IDENTITY OF THE HAZARDOUS CHEMICAL
- HAZARDOUS INGREDIENTS
- PHYSICAL AND CHEMICAL CHARACTERISTICS
- PHYSICAL HAZARDS
- HEALTH HAZARDS
- ROUTES OF ENTRY
- OSHA PERMISSIBLE ESPOSURE LIMIT
- CARCINOGEN INFORMATION
- CONTROL MEASURES
- EMERGENCY AND FIRST-AID PROCEDURES
- DATE OF PREPARATION AND NAME, ADDRESS AND TELEPHONE NUMBER OF RESPONSIBLE PARTY

## WRITTEN PROGRAM

A written hazard communication program has been developed and describes how the Hazard Communication Standard has been implemented in the workplace. It includes the following:

- CONTAINER LABELING PROVISIONS
- MSDS PROVISIONS
- A LIST OF HAZARDOUS CHEMICALS
- METHODS TO INFORM OF NON-ROUTINE TASKS
- PERSONS RESPONSIBLE FOR VARIOUS PROVISIONS
- LOCATION OF WRITTEN MATERIAL
- PROCEDURES AT MULTIEMPLOYEE WORKPLACES

THIS WRITTEN PROGRAM IS AVAILABLE TO EMPLOYEES UPON REQUEST

## CONTAINER LABELING

## IN-HOUSE LABELING SYSTEM

Standardized labeling, a numeric hazard rating of 0-4.

| | |
|---|---|
| H= Health | 4= Severe Hazard |
| F= Flammability | 3= Serious Hazard |
| R= Reactivity | 2= Moderate Hazard |
| PE= Protective Equipment | 1= Slight Hazard |
| | 0= Minimal Hazard |

Labels on hazardous chemicals provide an immediate warning to employees about the hazards of the chemicals.

CONSULT YOUR SUPERVISOR FOR MORE INFORMATION ABOUT THE CHEMICALS YOU WORK WITH.

**Figure 3-4: "Right-to-Know" Poster**

# Function Specifics[1]

## I. General

Specific safety awareness that is department or shop specific will be covered on the job by the department or shop supervisor. Each department or shop will warn of these hazards by visual means as well (e.g. signs or labels).

Every employee should be aware of the different methods of hazard recognition such as odor, liquid appearance, and the presence of warning signs. Employees should become familiar with their work area and the chemical products with which they work, as well as those in surrounding areas.

Employees should be aware of emergency procedures as established by company policy for response or evacuation purposes. Chemical fires should be handled with extreme caution.[2] Training and awareness of these subjects is accomplished through ***The Company's*** established Contingency Plan.

## II. Storage of Hazardous Materials

It is recommended that all hazardous materials be stored in a generalized location. Those materials that have a flammable rating of two (2) or higher should be stored in flammable storage cabinets. Some local governments have stricter storage requirements. This stricter requirement shall be observed (e.g. a local fire marshal may require that materials with any flammable rating be stored in flammable storage cabinets).

All cabinets and shelves should be organized in a manner so that like items are stored in adjacent locations.

---

[1] In this section, customer information can be addressed in addition to the requirements of the standard. For instance, if your company has a shelf-life policy for paints or adhesives, you can include information that the Receiving Department may be tasked to complete for company policy compliance.

[2] Your company should have an established contingency plan that covers fire response to include chemical fires. Although this program describes spill response, the contingency plan should cover this subject in greater detail to include facility or chemical specifics.

## Safety Awareness

Storage and handling of hazardous materials (or materials handling) offers many risks to the employee, some of which may not be immediately recognized by the employee. For example, an employee may be exposed to an unrealized risk to his/her health by transporting a chemical via forklift without first knowing the contents of the drum or consulting the appropriate MSDS. Any employee that handles hazardous materials should be aware of the material they are handling and be comfortable in knowing emergency procedures such as cleanup and first aid.

All employees should be aware of other safety precautions, such as the location of fire extinguishers and fire exits. Employees should be able to properly use transport equipment (e.g. must be trained and certified to operate the forklift) as well as fire extinguishing equipment. Furthermore, the employee must be alert to the dangers at hand and observe the work area to detect the presence or release of a hazardous chemical by visual appearance (noticing a leaky container) or the odor of the chemical when being released.

Take note of any product that enters your work area that you have never seen. The Program Coordinator should approve any new products that are purchased and/or introduced to your work area. Your supervisor or a trainer (depending on diversity of usage of the product) should demonstrate proper usage and safety precautions of any new product. If you find a new product for which you have not received any such information, notify your supervisor immediately. It is likely that your supervisor is unaware of the presence of the product.

Finally, all employees should read and understand the Hazard Communication Program including the storage and labeling requirements.

# Test

The following test will determine your knowledge of the HazCom Program and company policy. After completing this test, the trainer will review the results with you and correct any missed answers to ensure your complete understanding of the material. If you are unsure about any of the information provided, please ask as many questions as necessary.

1. This training qualifies the employee to ship or offer for shipment a package containing hazardous materials. True or False?

    a. True

    b. False

2. Chemical exposure may cause or contribute to:

    a. many serious health effects

    b. safety hazards

    c. fires

    d. all of the above

3. MSDS is an acronym for ______________________________.

4. One definition of a hazard is a possible source of danger. True or False?

    a. True

    b. False

5. Section I of an MSDS includes the Chemical Identification. The identity of the material must indicate the specific chemical name, common name, and proper shipping name of the chemical. True or False?

    a. True

    b. False

6. An acute health hazard is one that may show up immediately after exposure. True or False?

    a. True

    b. False

7. The Workplace Chemical List ________________ MSDSs in the workplace.

    a. replaces

    b. is a quick reference for

    c. shows locations for

8. The Hazardous Materials Identification System (HMIS) used by ***The Company*** communicates hazards with the following color codes:

   _____ Red a. Health

   _____ Yellow b. Flammability

   _____ Blue c. Reactivity

   _____ White d. Personal Protection required

9. The number four (4) in the flammability section of the Hazardous Materials Identification Guide (HMIG) indicates the item is:

   a. not flammable

   b. not likely to produce flame unless preheated

   c. slightly flammable with moderate heating

   d. moderately flammable

   e. extremely flammable

10. It is the responsibility of each employee to be aware of safety hazards and detect the presence or release of a hazardous substance. True or False?

    a. True

    b. False

11. It is the responsibility of the _________________ to assure that an MSDS accompanies each first-time shipment of a hazardous material.

    a. Employer

    b. Receiving Agent

    c. Chemical Manufacturer

    d. Program Coordinator

12. The HazCom Program that is most accessible to me is located ______________________.

13. New or revised MSDSs are sent to ____________________________ to be assigned a hazard code and added to the Workplace Chemical List.

14. If the personal protection index code is "D", you should wear what PPE when using the product?

    ____________________________________

15. Name three ways you can detect the presence of a hazardous chemical in your work area.

    1. ___________________________

    2. ___________________________

    3. ___________________________

16. A product with a flash point of 85°F should be assigned a _____ fire hazard rating.

## Matching

_____ 1. Corrosive Material

_____ 2. HMIS

_____ 3. DOT

_____ 4. Flash Point

_____ 5. Acute Hazard

_____ 6. Chronic Hazard

_____ 7. Carcinogen

_____ 8. Hazard

_____ 9. Risk

a. Substance or agent capable of causing cancer

b. Causes visible destruction or irreversible alterations in human skin tissue on contact

c. Minimum temperature at which a liquid gives off a sufficient concentration of vapors to ignite

d. Hazardous Materials Identification System

e. Long-term effects

f. Short-term effects

g. Possibility of suffering loss or harm

h. Department of Transportation

i. A chance happening

# Answers

1. b
2. d
3. Material Safety Data Sheet
4. a
5. b
6. a
7. b
8. b, c, a, d
9. e
10. a
11. c
12. (applicable location)
13. the Program Coordinator
14. Face shield, chemical gloves, and synthetic apron
15. Odor, Visually, Presence of signs/labels
16. 3

**Matching**

1. b
2. d
3. h
4. c
5. f
6. e
7. a
8. i
9. g

# Part VI

# Specific Requirements for Chemical and Physical Hazards

In this part are 16 sections that describe the requirements for specific chemical and physical hazards. They are offered in alphabetical order.

| | |
|---|---|
| Section 1 | 1, 3 - Butadiene (1910.1051) |
| Section 2 | 4 - Nitrobiphenyl (1910.1003) |
| Section 3 | Acrylonitrile (1910.1045) |
| Section 4 | Air Contaminants (1910.1000) |
| Section 5 | Asbestos (1910.1001) |
| Section 6 | Benzene (1910.1028) |
| Section 7 | Cadmium (1910.1027) |
| Section 8 | Cotton Dust (1910.1043) |
| Section 9 | 1,2-dibromo-3-chloropropane (DBCP) (1910.1044) |
| Section 10 | Ethylene Oxide (1910.1047) |
| Section 11 | Formaldehyde (1910.1048) |
| Section 12 | Inorganic Arsenic (1910.1018) |
| Section 13 | Lead (1910.1025) |
| Section 14 | Methylene Chloride (1910.1052) |
| Section 15 | Methylenedianiline (1910.1050) |
| Section 16 | Vinyl Chloride (1910.1017) |

These chemicals reflect those workplace chemicals that OSHA regulates specifically. In order to apply this information effectively, each reader will have an understanding of the chemical hazards present in the workplace through application of this book and OSHA's Hazard Communication Standard, 29 CFR 1910.1200. In no way should the reader expect to be required to comply with the contents of each section if that chemical is not present in the workplace. Each section should be considered applicable when that chemical or family of chemicals is, has been, or expects to be used in the workplace.

# 1, 3-Butadiene

## 1910.1051

Section 1 applies to all occupational exposures to 1,3-Butadiene (BD), Chemical Abstracts Service Registry No. 106-99-0.

## Definitions

- *Action level* means a concentration of airborne 1,3-Butadiene (BD) of 0.5 parts per million (ppm) calculated as an 8-hour time-weighted average (TWA).
- *Assistant Secretary* means the Assistant Secretary of Labor for Occupational Safety and Health, U.S. Department of Labor, or designee.
- *Authorized person* means any person specifically designated by the employer, whose duties require entrance into a regulated area, or a person entering such an area as a designated representative of employees to exercise the right to observe monitoring and measuring procedures.
- *1,3-Butadiene* means an organic compound with chemical formula $CH_2=CH\text{-}CH=CH_2$ that has a molecular weight of approximately 54.15 gm/mole.
- *Business day* means any Monday through Friday, except those days designated as federal, state, local, or company-specific holidays.
- *Complete Blood Count* (CBC) means laboratory tests performed on whole blood specimens and includes the following: white blood cell count (WBC), hematocrit (Hct), red blood cell count (RBC), hemoglobin (Hgb), differential count of white blood cells, red blood cell morphology, red blood cell indices, and platelet count.
- *Day* means any part of a calendar day.

- *Director* means the Director of the National Institute for Occupational Safety and Health (NIOSH), U.S. Department of Health and Human Services, or designee.

- *Emergency situation* means any occurrence such as, but not limited to, equipment failure, rupture of containers, or failure of control equipment that may or does result in an uncontrolled significant release of BD.

- *Employee exposure* means exposure of a worker to airborne concentrations of BD that would occur if the employee were not using respiratory protective equipment.

- *Objective data* means monitoring data, or mathematical modeling or calculations based on composition, chemical, and physical properties of a material, stream, or product.

- *Permissible Exposure Limits* (PELs) mean either the 8-hour TWA exposure or the Short-Term Exposure Limit (STEL).

- *Physician* or other licensed health care professional is an individual legally permitted scope of practice (i.e., license, registration, or certification) allows him or her to independently provide or be delegated the responsibility to provide one or more of the specific health care services.

- *Regulated area* means any area where airborne concentrations of BD exceed or can reasonably be expected to exceed the 8-hour TWA exposure of 1 ppm or the STEL of 5 ppm for 15 minutes.

- The *section* means this 1,3-butadiene standard.

## Permissible Exposure Limits (PELs)

- The employer shall ensure that no employee is exposed to an airborne concentration of BD in excess of 1 part BD per million parts of air (ppm) measured as an 8-hour TWA.

- The employer shall ensure that no employee is exposed to an airborne concentration of BD in excess of 5 ppm as determined over a sampling period of 15 minutes.

## Exposure Monitoring

- Determinations of employee exposure shall be made from breathing zone air samples that are representative of the 8-hour TWA and 15-minute short-term exposures of each employee.

- Representative 8-hour TWA employee exposure shall be determined on the basis of one or more samples representing full-shift exposure for each shift and for each job classification in each work area.

- Representative 15-minute short-term employee exposures shall be determined on the basis of one or more samples representing 15-minute exposures associated with operations that are most likely to produce exposures above the STEL for each shift and for each job classification in each work area.

- Except for the initial monitoring, where the employer can document that exposure levels are equivalent for similar operations on different work shifts, the employer need only determine representative employee exposure for that operation from the shift during which the highest exposure is expected.

- Each employer who has a workplace or work operation covered by this section, shall perform initial monitoring to determine accurately the airborne concentrations of BD to which employees may be exposed, or shall rely on objective data to fulfill this requirement.

- The employer may rely on such earlier monitoring results, provided that the conditions under which the initial monitoring was conducted have not changed in a manner that may result in new or additional exposures.

- If the initial monitoring reveals employee exposure to be at or above the action level but at or below both the 8-hour TWA limit and the STEL, the employer shall repeat the representative monitoring every 12 months.

- If the initial monitoring reveals employee exposure to be above the 8-hour TWA limit, the employer shall repeat the representative monitoring at least every three months until the employer has collected two samples per quarter (each at least seven days apart) within a two-year period, after which such monitoring must occur at least every six months.

- If the initial monitoring reveals employee exposure to be above the STEL, the employer shall repeat the representative monitoring at least every three months until the employer has collected two samples per quarter (each at least seven days apart) within a two-year period, after which such monitoring must occur at least every six months.

- The employer may alter the monitoring schedule from every six months to annually for any required representative monitoring for which two consecutive measurements taken at least seven days apart indicate that employee exposure has decreased to or below the 8-hour TWA, but is at or above the action level.

- If the initial monitoring reveals employee exposure to be below the action level and at or below the STEL, the employer may discontinue the monitoring for employees whose exposures are represented by the initial monitoring.

- If the periodic monitoring reveals that employee exposures, as indicated by at least two consecutive measurements taken at least seven days apart, are below the action level and at or below the STEL, the employer may discontinue the monitoring for those employees who are represented by such monitoring.

- The employer shall institute the exposure monitoring whenever there has been a change in the production, process, control equipment, personnel, or work practices that may result in new or additional exposures to BD or when the employer has any reason to suspect that a change may result in new or additional exposures.

- Whenever spills, leaks, ruptures, or other breakdowns occur that may lead to employee exposure above the 8-hour TWA limit or above the STEL, the employer shall monitor [using leak source, such as direct reading instruments, area, or personal monitoring] after the cleanup of the spill; or repair of the leak, rupture, or other breakdown; to ensure that exposures have returned to the level that existed prior to the incident.

- ❑ Monitoring shall be accurate, at a confidence level of 95 percent, to within plus or minus 25 percent for airborne concentrations of BD at or above the 1 ppm TWA limit and to within plus or minus 35 percent for airborne concentrations of BD at or above the action level of 0.5 ppm and below the 1 ppm TWA limit.

- ❑ The employer shall, within five business days after the receipt of the results of any monitoring performed under this section, notify the affected employees of these results in writing either individually or by posting results in an appropriate location that is accessible to all affected employees.

- ❑ The employer shall, within 15 business days after receipt of any monitoring performed indicating the 8-hour TWA or STEL has been exceeded, provide the affected employees, in writing, with information on the corrective action being taken by the employer to reduce employee exposure to or below the 8-hour TWA or STEL and the schedule for completion of this action.

- ❑ The employer shall provide affected employees or their designated representatives an opportunity to observe any monitoring of employee exposure to BD.

- ❑ When observation of the monitoring of employee exposure to BD requires entry into an area where the use of protective clothing or equipment is required, the employer shall at no cost provide the observer with protective clothing and equipment, and shall ensure that the observer uses this equipment and complies with all other applicable safety and health procedures.

## Regulated Areas

- ❑ The employer shall establish a regulated area wherever occupational exposures to airborne concentrations of BD exceed, or can reasonably be expected to exceed, the permissible exposure limits, either the 8-hour TWA or the STEL.

- ❑ Access to regulated areas shall be limited to authorized persons.

- ❑ Regulated areas shall be demarcated from the rest of the workplace in any manner that minimizes the number of employees exposed to BD within the regulated area.

- ❑ An employer at a multi-employer worksite who establishes a regulated area shall communicate the access restrictions and locations of these areas to other employers with work operations at that worksite whose employees may have access to these areas.

## Methods of Compliance

- ❑ The employer shall institute engineering controls and work practices to reduce and maintain employee exposure to or below the PELs, except to the extent that the employer can establish that these controls are not feasible.

- Wherever the feasible engineering controls and work practices which can be instituted are not sufficient to reduce employee exposure to or below the 8-hour TWA or STEL, the employer shall use them to reduce employee exposure to the lowest levels achievable by these controls and shall supplement them by the use of respiratory protection.

- Where any exposures are over the PELs, the employer shall establish and implement a written plan to reduce employee exposure to or below the PELs primarily by means of engineering and work practice controls, and by the use of respiratory protection. No compliance plan is required if all exposures are under the PELs.

- The written compliance plan shall include a schedule for the development and implementation of the engineering controls and work-practice controls, including periodic leak detection surveys.

- Copies of the compliance plan shall be furnished upon request for examination and copying to the Assistant Secretary, the Director, affected employees, and designated employee representatives. Such plans shall be reviewed at least every 12 months, and shall be updated as necessary to reflect significant changes in the status of the employer's compliance program.

- The employer shall not implement a schedule of employee rotation as a means of compliance with the PELs.

## Exposure Goal Program

- For those operations and job classifications where employee exposures are greater than the action level, in addition to compliance with the PELs, the employer shall have an exposure goal program that is intended to limit employee exposures to below the action level during normal operations.

- Written plans for the exposure goal program shall be furnished upon request for examination and copying to the Assistant Secretary, the Director, affected employees, and designated employee representatives.

- Such plans shall be updated as necessary to reflect significant changes in the status of the exposure goal program.

- Respirator use is not required in the exposure goal program.

- The exposure goal program shall include the following items unless the employer can demonstrate that the item is not feasible, will have no significant effect in reducing employee exposures, or is not necessary to achieve exposures below the action level:
    - A leak prevention, detection, and repair program.
    - A program for maintaining the effectiveness of local exhaust ventilation systems.
    - The use of pump exposure control technology such as, but not limited to, mechanical double-sealed or sealless pumps.

- Gauging devices designed to limit employee exposure, such as magnetic gauges on rail cars.
- Unloading devices designed to limit employee exposure, such as a vapor return system.
- A program to maintain BD concentration below the action level in control rooms by use of engineering controls.

## Respiratory Protection

- ❑ Respirators must be used during:
  - Periods necessary to install or implement feasible engineering and work-practice controls.
  - Non-routine work operations that are performed infrequently and for which employee exposures are limited in duration.
  - Work operations for which feasible engineering and work-practice controls are not yet sufficient to reduce employee exposures to or below the PELs.
  - Emergencies.
- ❑ The employer must implement a respiratory protection program in accordance with 29 CFR 1910.134 (b) through (d) (except (d)(1)(iii), and (d)(3)(iii)(B)(1) and (2)), and (f) through (m).
- ❑ If air-purifying respirators are used, the employer must replace the air-purifying filter elements according to the replacement schedule set for the class of respirators listed in Table 4-1-1 of this section, and at the beginning of each work shift.
- ❑ Instead of using the replacement schedule listed in Table 4-1-1 of this section, the employer may replace cartridges or canisters at 90 percent of their expiration service life, provided the employer:
  - Demonstrates that employees will be adequately protected by this procedure.
  - Uses BD breakthrough data for this purpose that have been derived from tests conducted under worst-case conditions of humidity, temperature, and air-flow rate through the filter element, and the employer also describes the data supporting the cartridge- or canister-change schedule, as well as the basis for using the data in the employer's respirator program.
- ❑ A label must be attached to each filter element to indicate the date and time it is first installed on the respirator.
- ❑ If NIOSH approves an end-of-service-life indicator (ESLI) for an air-purifying filter element, the element may be used until the ESLI shows no further useful service life or until the element is replaced at the beginning of the next work shift, whichever occurs first.
- ❑ Regardless of the air-purifying element used, if an employee detects the odor of BD, the employer must replace the air-purifying element immediately.
- ❑ The employer must select appropriate respirators from Table 4-1-1 of this section.

- Air-purifying respirators must have filter elements approved by NIOSH for organic vapors or BD.

- When an employee whose job requires the use of a respirator cannot use a negative-pressure respirator, the employer must provide the employee with a respirator that has less breathing resistance than the negative-pressure respirator, such as a powered air-purifying respirator or supplied-air respirator, when the employee is able to use it and if it provides the employee adequate protection.

- Where appropriate to prevent eye contact and limit dermal exposure to BD, the employer shall provide protective clothing and equipment at no cost to the employee and shall ensure its use. Eye and face protection shall meet the requirements of 29 CFR 1910.133.

## Table 4-1-1: Respiratory Protection for Airborne BD

| Concentration of airborne BD (ppm) or condition of use | Required respirator |
|---|---|
| Less than or equal to 5 ppm (5 X PEL) | Air-purifying half mask or full facepiece respirator equipped with approved BD or organic vapor cartridges or canisters. Cartridges or canisters shall be replaced every 4 hours. |
| Less than or equal to 10 ppm (10 X PEL) | Air-purifying half mask or full facepiece respirator equipped with approved BD or organic vapor cartridges or canisters. Cartridges or canisters shall be replaced every 3 hours. |
| Less than or equal to 25 ppm (25 X PEL) | Air-purifying full facepiece respirator equipped with approved BD or organic vapor cartridges or canisters. Cartridges or canisters shall be replaced every 2 hours; or<br>Powered air-purifying respirator equipped with approved BD or organic vapor cartridges. PAPR cartridges shall be replaced every 2 hours; or<br>Continuous flow supplied-air respirator equipped with a hood or helmet. |
| Less than or equal to 50 ppm (50 X PEL) | Air-purifying full facepiece respirator equipped with approved BD or organic vapor cartridges or canisters. Cartridges or canisters shall be replaced every hour; or<br>Powered air-purifying respirator equipped with a tight-fitting facepiece and an approved BD or organic vapor cartridges. PAPR cartridges shall be replaced every hour; or |
| Less than or equal to 1,000 ppm (1,000 X PEL) | Supplied-air respirator equipped with a half mask or full facepiece and operated in a pressure demand or other positive pressure mode. |
| Greater than 1,000 ppm, unknown concentration, or firefighting | Self-contained breathing apparatus equipped with a full facepiece and operated in a pressure demand or other positive pressure mode; or<br>Any supplied-air respirator equipped with a full facepiece and operated in a pressure demand or other positive pressure mode in combination with an auxiliary self-contained breathing apparatus operated in a pressure demand or other positive pressure mode. |
| Escape from IDLH conditions | Any positive pressure self-contained breathing apparatus with an appropriate service life; or<br>An air-purifying full facepiece respirator equipped with a front or back-mounted BD or organic vapor canister. |

Notes: Respirators approved for use in higher concentrations are permitted to be used in lower concentrations. Full facepiece is required when eye irritation is anticipated.

## Emergency Situations

- For each workplace where there is a possibility of an emergency, a written plan for emergency situations shall be developed, or an existing plan shall be modified, to contain the applicable elements specified in 29 CFR 1910.38, "Employee Emergency Plans and Fire Prevention Plans," and in 29 CFR 1910.120, "Hazardous Waste Operations and Emergency Responses."

## Medical Screening and Surveillance

- The employer shall institute a medical screening and surveillance program for:
  - Each employee with exposure to BD at concentrations at or above the action level on 30 or more days.
  - Employees who have or may have exposure to BD at or above the PELs on ten or more days a year;
- Employers (including successor owners) shall continue to provide medical screening and surveillance for each employee exposed to BD following an emergency situation and employees, even after transfer to a non-BD exposed job and regardless of when the employee is transferred, whose work histories suggest exposure to BD:
  - At or above the PELs on 30 or more days a year for ten or more years.
  - At or above the action level on 60 or more days a year for ten or more years.
  - Above 10 ppm on 30 or more days in any past year.
- The employer shall ensure that the health questionnaire, physical examination, and medical procedures are provided without cost to the employee, without loss of pay, and at a reasonable time and place.
- Physical examinations, health questionnaires, and medical procedures shall be performed or administered by a physician or other licensed health care professional.
- Laboratory tests shall be conducted by an accredited laboratory.
- For each employee who must wear a respirator, physical ability to perform the work and use the respirator must be determined as required by 29 CFR 1910.134.
- For each employee, a health questionnaire and complete blood count (CBC) with differential and platelet count shall be conducted every year. A physical examination shall be conducted as specified below:
  - An initial physical examination, if 12 months or more have elapsed since the last physical examination conducted as part of a medical screening program for BD exposure;
  - Before assumption of duties by the employee in a job with BD exposure;
  - Every three years after the initial physical examination;
  - At the discretion of the physician or other licensed health care professional reviewing the annual health questionnaire and CBC;

- At the time of employee reassignment to an area where exposure to BD is below the action level for continued coverage in the screening and surveillance program, and if 12 months or more have elapsed since the last physical examination; and
- At termination of employment if 12 months or more have elapsed since the last physical examination.

❑ Following an emergency situation, medical screening shall be conducted as quickly as possible, but not later than 48 hours after the exposure.

❑ Medical screening for employees shall include:

- A baseline health questionnaire that includes a comprehensive occupational and health history and is updated annually. Particular emphasis shall be placed on the hematopoietic and reticuloendothelial systems, including exposure to chemicals, in addition to BD, that may have an adverse effect on these systems, the presence of signs and symptoms that might be related to disorders of these systems, and any other information determined by the examining physician or other licensed health care professional to be necessary to evaluate whether the employee is at increased risk of material impairment of health from BD exposure.
- A complete physical examination, with special emphasis on the liver, spleen, lymph nodes, and skin.
- A CBC.
- Any other test which the examining physician or other licensed health care professional deems necessary to evaluate whether the employee may be at increased risk from exposure to BD.

❑ Medical screening for employees exposed to BD in an emergency situation shall focus on the acute effects of BD exposure and at a minimum include: a CBC within 48 hours of the exposure and then monthly for three months, and a physical examination if the employee reports blurred vision, coughing, drowsiness, nausea, headache, or irritation of the eyes, nose, throat, lungs, or skin. Continued employee participation in the medical screening and surveillance program, beyond these minimum requirements, shall be at the discretion of the physician or other licensed health care professional.

❑ Where the results of medical screening indicate abnormalities of the hematopoietic or reticuloendothelial systems for which a non-occupational cause is not readily apparent, the examining physician or other licensed health care professional shall refer the employee to an appropriate specialist for further evaluation and shall make available to the specialist the results of the medical screening.

❑ The specialist to whom the employee is referred shall determine the appropriate content for the medical evaluation (e.g., examinations, diagnostic tests and procedures, etc.).

❑ The employer shall provide the following information to the examining physician or other licensed health care professional involved in the evaluation:

- A copy of 29 CFR 1910.1051 including its appendices;

- A description of the affected employee's duties as they relate to the employee's BD exposure;
- The employee's actual or representative BD exposure level during employment tenure, including exposure incurred in an emergency situation;
- A description of pertinent personal protective equipment used or to be used; and
- Information, when available, from previous employment-related medical evaluations of the affected employee which is not otherwise available to the physician or other licensed health care professional or the specialist.

❑ For each medical evaluation required by this section, the employer shall ensure that the physician or other licensed health care professional produces a written opinion and provides a copy to the employer and the employee within 15 business days of the evaluation. The written opinion shall be limited to the following information:

- The occupationally pertinent results of the medical evaluation;
- A medical opinion concerning whether the employee has any detected medical conditions which would place the employee's health at increased risk of material impairment from exposure to BD;
- Any recommended limitations upon the employee's exposure to BD; and
- A statement that the employee has been informed of the results of the medical evaluation and any medical conditions resulting from BD exposure that require further explanation or treatment.

❑ The written medical opinion provided to the employer shall not reveal specific records, findings, and diagnoses that have no bearing on the employee's ability to work with BD.

> Note: *This provision does not negate the ethical obligation of the physician or other licensed health care professional to transmit any other adverse findings directly to the employee.*

❑ The employer shall ensure that information obtained from the medical screening program activities is aggregated (with all personal identifiers removed) and periodically reviewed to ascertain whether the health of the employee population of that employer is adversely affected by exposure to BD.

❑ Information learned from medical surveillance activities must be disseminated to covered employees in a manner that ensures the confidentiality of individual medical information.

## Communication of Hazards to Employees

❑ The employer shall communicate the hazards associated with BD exposure in accordance with the requirements of the Hazard Communication Standard, 29 CFR 1910.1200, 29 CFR 1915.1200, and 29 CFR 1926.59.

❑ The employer shall provide all employees exposed to BD with information and training in accordance with the requirements of the Hazard Communication Standard, 29 CFR 1910.1200, 29 CFR 1915.1200, and 29 CFR 1926.59.

- ❑ The employer shall institute a training program for all employees who are potentially exposed to BD at or above the action level or the STEL, ensure employee participation in the program, and maintain a record of the contents of such a program.

- ❑ Training shall be provided prior to, or at the time of, initial assignment to a job potentially involving exposure to BD at or above the action level or STEL, and at least annually thereafter.

- ❑ The training program shall be conducted in a manner that the employee is able to understand. The employer shall ensure that each employee exposed to BD over the action level or STEL is informed of the following:
  - The health hazards associated with BD exposure, and the purpose and a description of the medical screening and surveillance program;
  - The quantity, location, manner of use, release, and storage of BD and the specific operations that could result in exposure to BD, especially exposures above the PEL or STEL;
  - The engineering controls and work practices associated with the employee's job assignment, emergency procedures, and personal protective equipment;
  - The measures employees can take to protect themselves from exposure to BD;
  - The contents of 29 CFR 1910.1051 and its appendices; and
  - The right of each employee exposed to BD at or above the action level or STEL to obtain:
    - Medical examinations at no cost to the employee;
    - The employee's medical records required to be maintained; and
    - All air monitoring results representing the employee's exposure to BD and required to be kept.

- ❑ The employer shall make a copy of 29 CFR 1910.1051 and its appendices readily available without cost to all affected employees and their designated representatives, and shall provide a copy if requested.

- ❑ The employer shall provide upon request to the Assistant Secretary or the Director, or the designated employee representatives, all materials relating to the employee information and training program.

## Recordkeeping

- ❑ Where the processing, use, or handling of products or streams made from or containing BD are exempted from other requirements, or where objective data have been relied upon in lieu of initial monitoring, the employer shall establish and maintain a record of the objective data reasonably relied upon in support of the exemption.

- ❑ This record shall include at least the following information:
  - The product or activity qualifying for exemption;
  - The source of the objective data;
  - The testing protocol, results of testing, and analysis of the material for the release of BD;

  - A description of the operation exempted and how the data support the exemption; and
  - Other data relevant to the operations, materials, processing, or employee exposures covered by the exemption.

- The employer shall maintain this record for the duration of the employer's reliance upon such objective data.

- The employer shall establish and maintain an accurate record of all measurements taken to monitor employee exposure to BD.

- The record shall include at least the following information:
  - The date of measurement;
  - The operation involving exposure to BD which is being monitored;
  - Sampling and analytical methods used and evidence of their accuracy;
  - Number, duration, and results of samples taken;
  - Type of protective devices worn, if any;
  - Name, social security number, and exposure of the employees whose exposures are represented; and
  - The written corrective action and the schedule for completion of this action.

- The employer shall maintain this record for at least 30 years in accordance with 29 CFR 1910.20.

- The employer shall establish a record of the fit tests administered to an employee including:
  - Name of the employee;
  - Type of respirator;
  - Brand and size of respirator;
  - Date of test; and
  - Where QNFT is used, the fit factor, strip chart recording, or other recording of the results of the test.

- Fit test records shall be maintained for respirator users until the next fit test is administered.

- The employer shall establish and maintain an accurate record for each employee subject to medical screening and surveillance.

- The record shall include at least the following information:
  - The name and social security number of the employee;
  - A physician's or other licensed health care professional's written opinions; and
  - A copy of the information provided to the physician or other licensed health care professional.

- Medical screening and surveillance records shall be maintained for each employee for the duration of employment plus 30 years, in accordance with 29 CFR 1910.20.

- ❑ The employer, upon written request, shall make all records available to the Assistant Secretary and the Director for examination and copying.
- ❑ Access to records to be maintained shall be granted in accordance with 29 CFR 1910.20(e).
- ❑ Whenever the employer ceases to do business, the employer shall transfer records to the successor employer. The successor employer shall receive and maintain these records. If there is no successor employer, the employer shall notify the Director, at least three months prior to disposal, and transmit the records to the Director if requested by the Director within that period.
- ❑ The employer shall transfer medical and exposure records as set forth in 29 CFR 1910.20(h).

# 4-Nitrobiphenyl

## 1910.1003

Section 2 applies to any area in which the 13 carcinogens addressed by this section are manufactured, processed, repackaged, released, handled, or stored. It does, however, apply only to shipments in sealed containers for the purpose of labeling requirements.

- The 13 carcinogens are:
  - 4-Nitrobiphenyl, Chemical Abstracts Service Register Number (CAS No.) 92933;
  - alpha-Naphthylamine, CAS No. 134327;
  - methyl chloromethyl ether, CAS No. 107302;
  - 3,'-Dichlorobenzidine (and its salts) CAS No. 91941;
  - bis-Chloromethyl ether, CAS No. 542881;
  - beta-Naphthylamine, CAS No. 91598;
  - Benzidine, CAS No. 92875;
  - 4-Aminodiphenyl, CAS No. 92671;
  - Ethyleneimine, CAS No. 151564;
  - beta-Propiolactone, CAS No. 57578;
  - 2-Acetylaminofluorene, CAS No. 53963;
  - 4-Dimethylaminoazo-benezene, CAS No. 60117; and
  - N-Nitrosodimethylamine, CAS No. 62759.
- This section shall not apply to the following:
  - Solid or liquid mixtures containing less than 0.1 percent by weight or volume of 4-Nitrobiphenyl, methyl chloromethyl ether, bis-chloromethyl ether, beta-Naphthylamine, benzidine, and 4-Aminodiphenyl; and

- Solid or liquid mixtures containing less than 1.0 percent by weight or volume of alpha-Naphthylamine, 3,'-Dichlorobenzidine (and its salts), Ethyleneimine, beta-Propiolactone, 2-Acetylaminofluorene, 4-Dimethylaminoazobenzene, or N-Nitrosodimethylamine.

## Definitions

- ❑ *Absolute filter* is one capable of retaining 99.97 percent of a mono disperse aerosol of 0.3 mm particles.
- ❑ *Authorized employee* means an employee whose duties require him to be in the regulated area and whom the employer has specifically assigned.
- ❑ *Clean change room* means a room where employees put on clean clothing and/or protective equipment in an environment free of the 13 carcinogens addressed by this section. The clean change room shall be contiguous to and have an entry from a shower room when shower room facilities are otherwise required.
- ❑ *Closed system* means an operation involving a carcinogen where containment prevents the release of the material into regulated areas, nonregulated areas, or the external environment.
- ❑ *Decontamination* means the inactivation of a carcinogen or its safe disposal.
- ❑ *Director* means the Director of National Institute for Occupational Safety and Health, or any person directed by him or the Secretary of Health and Human Services to act for the Director.
- ❑ *Disposal* means the safe removal of the carcinogens from the work environment.
- ❑ *Emergency* means an unforeseen circumstance or set of circumstances resulting in the release of a carcinogen that may result in exposure to or contact with the material.
- ❑ *External environment* means any environment external to regulated and nonregulated areas.
- ❑ *Isolated system* means a fully enclosed structure other than the vessel of containment of a carcinogen that is impervious to the passage of the material and would prevent the entry of the carcinogen into regulated areas, nonregulated areas, or the external environment, should leakage or spillage from the vessel of containment occur.
- ❑ *Laboratory-type hood* means a device enclosed on the three sides and the top and bottom, designed and maintained so as to draw air inward at an average linear face velocity of 150 feet per minute with a minimum of 125 feet per minute; and designed, constructed, and maintained in such a way that an operation involving a carcinogen within the hood does not require the insertion of any portion of any employee's body other than his hands and arms.
- ❑ *Nonregulated area* means any area under the control of the employer where entry and exit is neither restricted nor controlled.
- ❑ *Open-vessel system* means an operation involving a carcinogen addressed by this section in an open vessel that is not in an isolated system, a laboratory-type hood, nor any other system

affording equivalent protection against the entry of the material into regulated areas, nonregulated areas, or the external environment.

- *Protective clothing* means clothing designed to protect an employee against contact with or exposure to a carcinogen.

- *Regulated area* means an area where entry and exit is restricted and controlled.

## General

- A regulated area shall be established by an employer where a carcinogen addressed by this section is manufactured, processed, used, repackaged, released, handled, or stored. All such areas shall be controlled in accordance with the requirements for the following category or categories describing the operation involved:
  - Employees working with a carcinogen addressed by this section within an isolated system such as a "glove box" shall wash their hands and arms upon completion of the assigned task and before engaging in other activities not associated with the isolated system.
  - Within regulated areas where the carcinogens are stored in sealed containers or contained in a closed system, including piping systems, with any sample ports or openings closed while the carcinogens addressed by this section are contained within, access shall be restricted to authorized employees only.

- Employees exposed to 4-Nitrobiphenyl; alpha-Naphthylamine; 3,'-Dichlorobenzidine (and its salts); betanisters or cartridges. A respirator affording higher levels of protection than this respirator may be substituted.

- Prior to each exit from a regulated area, employees shall be required to remove and leave protective clothing and equipment at the point of exit. At the last exit of the day, employees shall place used clothing and equipment in impervious containers at the point of exit for purposes of decontamination or disposal.

- Drinking fountains are prohibited in the regulated area.

- Employees shall be required to wash hands, forearms, face, and neck on each exit from the regulated area, close to the point of exit, and before engaging in other activities. Employees exposed to 4-Nitrobiphenyl; alpha-Naphthylamine; 3,'-Dichlorobenzidine (and its salts); beta-Naphthylamine; Benzidine; 4-Aminodiphenyl; 2-Acetylaminofluorene; 4-Dimethylaminoazobenzene; or N-Nitrosodimethylamine shall be required to shower after the last exit of the day.

## Maintenance and Cleanup

- In cleanup of leaks or spills, maintenance, or repair operations on contaminated systems or equipment, or any operations involving work in an area where direct contact with a carcinogen could result, each authorized employee entering that area shall:

- Be provided with and required to wear clean, impervious garments, including gloves, boots, and continuous air-supplied hood in accordance with 29 CFR 1910.134.
- Be decontaminated before removing the protective garments and hood.
- Be required to shower upon removing the protective garments and hood.

## Respiratory Protection

- ❑ The employer must implement a respiratory protection program in accordance with 29 CFR 1910.134 (b), (c), (d) (except (d)(1)(iii) and (iv) and (d)(3)), and (e) through (m).

## Emergency Situations

- ❑ In an emergency, immediate measures of this section shall be implemented.
- ❑ The potentially affected area shall be evacuated as soon as the emergency has been determined.
- ❑ Hazardous conditions created by the emergency shall be eliminated and the potentially affected area shall be decontaminated prior to the resumption of normal operations.
- ❑ Special medical surveillance by a physician shall be instituted within 24 hours for employees present in the potentially affected area at the time of the emergency. A report of the medical surveillance and any treatment shall be included in the incident report.
- ❑ Where an employee has a known contact with a carcinogen addressed by this section, such employee shall be required to shower as soon as possible, unless contraindicated by physical injuries.
- ❑ An incident report on the emergency shall be reported.
- ❑ Emergency deluge showers and eyewash fountains supplied with running potable water shall be located near, within sight of, and on the same level as locations where a direct exposure to Ethyleneimine or beta-Propiolactone would be most likely only as a result of equipment failure or improper work practice.

## Hygiene Facilities and Practices

- ❑ Storage or consumption of food; storage or use of containers of beverages; storage or application of cosmetics; smoking; storage of smoking materials, tobacco products, or other products for chewing; or the chewing of such products are prohibited in regulated areas.
- ❑ Where employees are required to wash, washing facilities shall be provided in accordance with 29 CFR 1910.141(d)(1) and (2)(ii) through (vii).
- ❑ Where employees are required to shower, shower facilities shall be provided in accordance with 29 CFR 1910.141(d)(3).

- Where employees wear protective clothing and equipment, clean change rooms shall be provided for the number of such employees required to change clothes in accordance with 29 CFR 1910.141(e).

- Where toilets are in regulated areas, such toilets shall be in a separate room.

## Contamination Control

- Except for outdoor systems, regulated areas shall be maintained under negative pressure with respect to nonregulated areas. Local exhaust ventilation may be used to satisfy this requirement. Clean makeup air in equal volume shall replace air removed.

- Any equipment, material, or other item taken into or removed from a regulated area shall be done so in a manner that does not cause contamination in nonregulated areas or the external environment.

- Decontamination procedures shall be established and implemented to remove carcinogens from the surfaces of materials, equipment, and the decontamination facility.

- Dry sweeping and dry mopping are prohibited for 4-Nitrobiphenyl; alpha-Naphthylamine; 3,'-Dichlorobenzidine (and its salts); beta-Naphthylamine; Benzidine; 4-Aminodiphenyl; 2-Acetylaminofluorene; 4-Dimethylaminoazo-benzene; and N-Nitrosodimethylamine.

## Communication of Hazards to Employees

- Entrances to regulated areas shall be posted with appropriate caution signs.

- Appropriate signs and instructions shall be posted at the entrance to and exit from regulated areas, informing employees of the procedures that must be followed in entering and leaving a regulated area.

- Containers of a carcinogen that are accessible only to, and handled only by, authorized employees or trained employees may have contents identification limited to a generic or proprietary name or other proprietary identification of the carcinogen and percent.

- Containers of carcinogen containers that are accessible to or handled by employees other than authorized employees or trained employees shall have content identification that includes the full chemical name and Chemical Abstracts Service Registry number.

- Containers shall have the warning words "CANCER-SUSPECT AGENT" displayed immediately under or adjacent to the contents identification.

- ❑ Containers whose contents are carcinogens with corrosive or irritating properties shall have label statements warning of such hazards noting, if appropriate, particularly sensitive or affected portions of the body.

- ❑ Lettering on signs and instructions shall be a minimum letter height of 2 inches (5 cm). Labels on containers shall not be less than one-half the size of the largest lettering on the package, and not less than 8-point type in any instance. Provided no such required lettering need be more than 1 inch (2.5 cm) in height.

- ❑ No statement shall appear on or near any required sign, label, or instruction that contradicts or detracts from the effect of any required warning, information, or instruction.

## Training

- ❑ Each employee prior to being authorized to enter a regulated area, shall receive a training and indoctrination program including, but not necessarily limited to:
    - The nature of the carcinogenic hazards of a carcinogen, including local and systemic toxicity;
    - The specific nature of the operation involving a carcinogen that could result in exposure;
    - The purpose for and application of the medical surveillance program, including, as appropriate, methods of self-examination;
    - The purpose for and application of decontamination practices and purposes;
    - The purpose for and significance of emergency practices and procedures;
    - The employee's specific role in emergency procedures;
    - Specific information to aid the employee in recognition and evaluation of conditions and situations which may result in the release of a carcinogen;
    - The purpose for, and application of, specific first aid procedures and practices; and
    - A review of the employee's first training and indoctrination program, and annually thereafter.

- ❑ Specific emergency procedures shall be prescribed, and posted, and employees shall be familiarized with their terms and rehearsed in their application.

- ❑ All materials relating to the program shall be provided upon request to authorized representatives of the Assistant Secretary and the Director.

## Reports

- ❑ The following information shall be reported in writing to the nearest OSHA Area Director, and any changes in such information shall be similarly reported in writing within 15 calendar days of such change:
    - A brief description and in-plant location of the area(s) regulated and the address of each regulated area;

- The name(s) and other identifying information as to the presence of a carcinogen in each regulated area;
- The number of employees in each regulated area during normal operations, including maintenance activities; and
- The manner in which carcinogens are present in each regulated area; for example, whether it is manufactured, processed, used, repackaged, released, stored, or otherwise handled.

## Incidents

- Incidents that result in the release of a carcinogen into any area where employees may be potentially exposed shall be reported.

- A report of the occurrence of the incident and the facts obtainable at that time, including a report on any medical treatment of affected employees, shall be made within 24 hours to the nearest OSHA Area Director.

- A written report shall be filed with the nearest OSHA Area Director within 15 calendar days thereafter and shall include:
  - A specification of the amount of material released, the amount of time involved, and an explanation of the procedure used in determining this figure;
  - A description of the area involved, and the extent of known and possible employee exposure and area contamination;
  - A report of any medical treatment of affected employees, and any medical surveillance program implemented; and
  - An analysis of the circumstances of the incident and measures taken or to be taken, with specific completion dates, to avoid further similar releases.

## Medical Surveillance

- At no cost to the employee, a program of medical surveillance shall be established and implemented for employees considered for assignment to enter regulated areas, and for authorized employees.

- Before an employee is assigned to enter a regulated area, a preassignment physical examination by a physician shall be provided. The examination shall include the personal history of the employee, and family and occupational background, including genetic and environmental factors.

- Authorized employees shall be provided periodic physical examinations, not less often than annually, following the preassignment examination.

- In all physical examinations, the examining physician shall consider whether there exist conditions of increased risk, including reduced immunological competence, treatment with steroids or cytotoxic agents, pregnancy, and cigarette smoking.

**Recordkeeping**

- ❑ Employers of all employees examined will maintain complete and accurate records of all medical examinations. Records shall be maintained for the duration of the employee's employment. Upon termination of the employee's employment, including retirement or death, or in the event that the employer ceases business without a successor, records, or notarized true copies thereof, shall be forwarded by registered mail to the Director.

- ❑ Records shall be provided upon request to employees, designated representatives, and the Assistant Secretary in accordance with 29 CFR 1910.20 (a) through (e) and (g) through (i). These records shall also be provided upon request to the Director.

- ❑ Any physician who conducts a medical examination shall furnish to the employer a statement of the employee's suitability for employment in the specific exposure.

# Acrylonitrile

## 1910.1045

Section 3 applies to all occupational exposures to acrylonitrile (AN), CAS No. 000107131. This section does not apply to exposures which result solely from the processing, use, and handling of ABS resins, SAN resins, nitrile barrier resins, solid nitrile elastomers, and acrylic and modacrylic fibers, when these listed materials are in the form of finished polymers, and products fabricated from such finished polymers. This section also does not apply to materials made from and/or containing AN for which objective data is reasonably relied upon to demonstrate that the material is not capable of releasing AN in airborne concentrations in excess of 1 ppm as an 8-hour time-weighted average (TWA), under the expected conditions of processing, use, and handling which will cause the greatest possible release.

## Definitions

- *Acrylonitrile* or AN means acrylonitrile monomer, chemical formula $CH_2=CHCN$.
- *Action level* means a concentration of AN of 1 ppm as an 8-hour TWA.
- *Assistant Secretary* means the Assistant Secretary of Labor for Occupational Safety and Health, U.S. Department of Labor, or designee.
- *Authorized person* means any person specifically authorized by the employer whose duties require the person to enter a regulated area, or any person entering such an area as a designated representative of employees for the purpose of exercising the opportunity to observe monitoring.
- An *emergency* is any occurrence such as, but not limited to, equipment failure, rupture of containers, or failure of control equipment which results in an unexpected massive release of AN.

- *Liquid AN* means AN monomer in liquid form, and liquid or semiliquid polymer intermediates, including slurries, suspensions, emulsions, and solutions, produced during the polymerization of AN.

- *OSHA Area Office* means the Area Office of the Occupational Safety and Health Administration having jurisdiction over the geographic area where the affected workplace is located.

## Permissible Exposure Limits (PELs)

- The employer shall assure that no employee is exposed to an airborne concentration of acrylonitrile in excess of 2 parts acrylonitrile per million parts of air (2 ppm) as an 8-hour TWA.

- The employer shall assure that no employee is exposed to an airborne concentration of acrylonitrile in excess of 10 ppm as averaged over any 15-minute period during the work day.

- The employer shall assure that no employee is exposed to skin contact or eye contact with liquid AN.

## Notification of Regulated Areas and Emergencies

- Within 30 days following the establishment of a regulated area, the employer shall report the following information to the OSHA Area Office:
  - The address and location of each establishment which has one or more regulated areas;
  - The locations, within the establishment, of each regulated area;
  - A brief description of each process or operation which results in employee exposure to AN in regulated areas; and
  - The number of employees engaged in each process or operation within each regulated area that results in exposure to AN, and an estimate of the frequency and degree of exposure that occurs.

- Whenever there has been a significant change in the information required to be reported, the employer shall promptly provide the new information to the OSHA Area Office.

- Emergencies, and the facts obtainable at that time, shall be reported to the OSHA Area Office within 72 hours of the initial occurrence. Upon request of the OSHA Area Office, the employer shall submit additional information in writing relevant to the nature and extent of employee exposures and measures taken to prevent future emergencies of a similar nature.

## Exposure Monitoring

- Determinations of airborne exposure levels shall be made from air samples that are representative of each employee's exposure to AN over an 8-hour period.

- Employee exposure is that exposure which would occur if the employee were not using a respirator.

- Each employer who has a place of employment in which AN is present shall monitor each such workplace and work operation to accurately determine the airborne concentrations of AN to which employees may be exposed.

- If the monitoring required by this section reveals employee exposure to be below the action level, the employer may discontinue monitoring for that employee.

- If the monitoring reveals employee exposure to be at or above the action level but below the permissible exposure limits, the employer shall repeat such monitoring for each such employee at least quarterly. The employer shall continue these quarterly measurements until at least two consecutive measurements taken at least seven days apart, are below the action level, and thereafter the employer may discontinue monitoring for that employee.

- If the monitoring reveals employee exposure to be in excess of the permissible exposure limits, the employer shall repeat these determinations for each such employee at least monthly. The employer shall continue these monthly measurements until at least two consecutive measurements, taken at least seven days apart, are below the permissible exposure limits, and thereafter the employer shall monitor at least quarterly.

- Whenever there has been a production, process, control, or personnel change which may result in new or additional exposures to AN, or whenever the employer has any other reason to suspect a change which may result in new or additional exposures to AN, additional monitoring shall be conducted.

- Within five working days after the receipt of the results of monitoring, the employer shall notify each employee in writing of the results which represent that employee's exposure.

- Whenever the results indicate that the representative employee exposure exceeds the permissible exposure limits, the employer shall include in the written notice a statement that the permissible exposure limits were exceeded and a description of the corrective action being taken to reduce exposure to or below the permissible exposure limits.

- The method of measurement of employee exposures shall be accurate to a confidence level of 95 percent, to within plus or minus 35 percent for concentrations of AN at or above the permissible exposure limits, and plus or minus 50 percent for concentrations of AN below the permissible exposure limits.

## Regulated Areas

- The employer shall establish regulated areas where AN concentrations are in excess of the permissible exposure limits.

- Regulated areas shall be demarcated and segregated from the rest of the workplace in any manner that minimizes the number of persons who will be exposed to AN.

- Access to regulated areas shall be limited to authorized persons.

- ❑ The employer shall assure that food or beverages are not present or consumed, tobacco products are not present or used, and cosmetics are not present or applied in the regulated area.

## Methods of Compliance

- ❑ The employer shall institute engineering and work-practice controls to reduce and maintain employee exposures to AN to or below the permissible exposure limits, except to the extent that the employer establishes that such controls are not feasible.

- ❑ Wherever the engineering and work-practice controls which can be instituted are not sufficient to reduce employee exposures to or below the permissible exposure limits, the employer will use them to reduce exposures to the lowest levels achievable by these controls, and shall supplement them by the use of respiratory protection.

- ❑ The employer shall establish and implement a written program to reduce employee exposures to or below the permissible exposure limits solely by means of engineering and work-practice controls.

- ❑ Written plans for these compliance programs shall include at least the following:
  - A description of each operation or process resulting in employee exposure to AN above the permissible exposure limits;
  - An outline of the nature of the engineering controls and work practices to be applied to the operation or process in question;
  - A report of the technology considered in meeting the permissible exposure limits; and
  - Other relevant information.

- ❑ The employer shall complete the steps set forth in the compliance program by the dates in the schedule.

- ❑ Written plans shall be submitted upon request to the Assistant Secretary and the Director, and shall be available at the worksite for examination and copying by the Assistant Secretary, the Director, or any affected employee or representative.

- ❑ The plans required by this paragraph shall be revised and updated at least every six months to reflect the current status of the program.

## Respiratory Protection

- ❑ The employer must provide respirators that comply with the requirements of this paragraph. Respirators must be used during:
  - Periods necessary to install or implement feasible engineering and work-practice controls.
  - Work operations, such as maintenance and repair activities or reactor cleaning, for which the employer establishes that engineering and work-practice controls are not feasible.

- Work operations for which feasible engineering and work-practice controls are not yet sufficient to reduce employee exposure to or below the permissible exposure limits.
- Emergencies.

❑ The employer must implement a respiratory protection program in accordance with 29 CFR 1910.134 (b) through (d) (except (d)(1)(iii) and (d)(3)(iii)(B)(1) and (2)), and (f) through (m).

❑ If air-purifying respirators (chemical-cartridge or chemical-canister types) are used:

- The air-purifying canister or cartridge must be replaced prior to the expiration of its service life or at the completion of each shift, whichever occurs first.
- A label must be attached to the cartridge or canister to indicate the date and time at which it is first installed on the respirator.

❑ The employer must select the appropriate respirator from Table 4-3-1.

## Table 4-3-1: Respiratory Protection for AN

| Concentration of AN or condition of use | Required respirator |
|---|---|
| Less than or equal to 20 ppm | Chemical cartridge respirator with organic vapor cartridge(s) and half-mask facepiece; or Supplied-air respirator with half-mask facepiece. |
| Less than or equal to 100 ppm or maximum use concentration (MUC) of cartridges or canisters, whichever is lower | Full facepiece respirator with (A) organic vapor cartridges; (B) organic vapor gas mask chin-style; or (C) organic vapor gas mask canister, front- or back-mounted; or<br>Supplied-air respirator with full facepiece; or<br>Self-contained breathing apparatus with full facepiece. |
| Less than or equal to 4,000 ppm | Supplied-air respirator operated in the positive-pressure mode with full facepiece, helmet, suit, or hood. |
| Greater than 4,000 ppm or unknown concentration | Supplied-air and auxiliary self-contained breathing apparatus with full facepiece in positive-pressure mode; or<br>Self-contained breathing apparatus with full facepiece in positive-pressure mode. |
| Firefighting | Self-contained breathing apparatus with full facepiece in positive-pressure mode. |
| Escape | Any organic vapor respirator, or<br>Any self-contained breathing apparatus. |

## Emergency Situations

- ❑ A written plan for emergency situations shall be developed for each workplace where liquid AN is present. Appropriate portions of the plan shall be implemented in the event of an emergency.
- ❑ The plan shall specifically provide that employees engaged in correcting emergency conditions shall be equipped until the emergency is abated.
- ❑ Employees not engaged in correcting the emergency shall be evacuated from the area and shall not be permitted to return until the emergency is abated.
- ❑ Where there is the possibility of employee exposure to AN in excess of the ceiling limit, a general alarm shall be installed and used to promptly alert employees of such occurrences.

## Protective Clothing and Equipment

- ❑ Where eye or skin contact with liquid AN may occur, the employer shall provide at no cost to the employee, and assure that employees wear, impermeable protective clothing or other equipment to protect any area of the body which may come in contact with liquid AN. The employer shall comply with the provision of 29 CFR 1910.132 and 1910.133.
- ❑ The employer shall clean, launder, maintain, or replace protective clothing and equipment as needed to maintain their effectiveness.
- ❑ The employer shall assure that impermeable protective clothing which contacts or is likely to have contacted liquid AN shall be decontaminated before being removed by the employee.
- ❑ The employer shall assure that an employee whose non-impermeable clothing becomes wetted with liquid AN shall immediately remove that clothing and proceed to the shower. The clothing shall be decontaminated before it is removed from the regulated area.
- ❑ The employer shall assure that no employee removes protective clothing or equipment from the change room, except for those employees authorized to do so for the purpose of laundering, maintenance, or disposal.
- ❑ The employer shall inform any person who launders or cleans protective clothing or equipment of the potentially harmful effects of exposure to AN.

## Housekeeping

- ❑ All surfaces shall be maintained free of visible accumulations of liquid AN.
- ❑ For operations involving liquid AN, the employer shall institute a program for detecting leaks and spills of liquid AN, including regular visual inspections.

- Where spills of liquid AN are detected, the employer shall assure that surfaces contacted by the liquid AN are decontaminated. Employees not engaged in decontamination activities shall leave the area of the spill, and shall not be permitted in the area until decontamination is completed.

## Waste Disposal

- AN waste, scrap, debris, bags, containers, or equipment shall be decontaminated before being incorporated in the general waste disposal system.

## Hygiene Facilities and Practices

- Where employees are exposed to airborne concentrations of AN above the permissible exposure limits, or where employees are required to wear protective clothing or equipment, the facilities required by 29 CFR 1910.141, including clean change rooms and shower facilities, shall be provided by the employer for the use of those employees.

- The employer shall assure that employees wearing protective clothing or equipment for protection of skin from liquid AN shall shower at the end of the work shift.

- The employer shall assure that, in the event of skin or eye exposure to liquid AN, the affected employee shall shower immediately to minimize the danger of skin absorption.

- The employer shall assure that employees working in the regulated area wash their hands and faces prior to eating.

## Medical Surveillance

- The employer shall institute a program of medical surveillance for each employee who is or will be exposed to AN at or above the action level, without regard to the use of respirators. The employer shall provide each such employee with an opportunity for medical examinations and tests.

- The employer shall assure that all medical examinations and procedures are performed by or under the supervision of a licensed physician, and that they shall be provided without cost to the employee.

- At the time of initial assignment, or upon institution of the medical surveillance program, the employer shall provide each affected employee an opportunity for a medical examination, including at least the following elements:

  - A work history and medical history with special attention to skin, respiratory, and gastrointestinal systems, and those nonspecific symptoms, such as headache, nausea, vomiting, dizziness, weakness, or other central nervous system dysfunctions that may be associated with acute or chronic exposure to AN;
  - A complete physical examination giving particular attention to the peripheral and central nervous system, gastrointestinal system, respiratory system, skin, and thyroid;

- A 14 by 17-inch posteroanterior chest X-ray; and
- Further tests of the intestinal tract, including fecal occult blood screening, for all workers 40 years of age or older, and for any other affected employees for whom, in the opinion of the physician, such testing is appropriate.

❑ The employer shall provide the examinations at least annually for employees.

❑ If an employee has not had the examination within six months preceding termination of employment, the employer shall make such examination available to the employee prior to such termination.

❑ If the employee for any reason develops signs or symptoms which may be associated with exposure to AN, the employer shall provide an appropriate examination and emergency medical treatment.

❑ The employer shall provide the following information to the examining physician:

- A copy of 29 CFR 1910.1045 and its appendixes;
- A description of the affected employee's duties as they relate to the employee's exposure;
- The employee's representative exposure level;
- The employee's anticipated or estimated exposure level (for preplacement examinations or in cases of exposure due to an emergency);
- A description of any personal protective equipment used or to be used; and
- Information from previous medical examinations of the affected employee, which is not otherwise available to the examining physician.

❑ The employer shall obtain a written opinion from the examining physician which shall include:

- The results of the medical examination and test performed;
- The physician's opinion as to whether the employee has any detected medical condition(s) which would place the employee at an increased risk of material impairment of the employee's health from exposure to AN;
- Any recommended limitations upon the employee's exposure to AN or upon the use of protective clothing and equipment such as respirators; and
- A statement that the employee has been informed by the physician of the results of the medical examination and any medical conditions which require further examination or treatment.

❑ The employer shall instruct the physician not to reveal in the written opinion specific findings or diagnoses unrelated to occupational exposure to AN.

❑ The employer shall provide a copy of the written opinion to the affected employee.

## Communication of Hazards to Employees

- ❑ The employer shall institute a training program for, and assure the participation of, all employees exposed to AN above the action level, all employees whose exposures are maintained below the action level by engineering and work-practice controls, and all employees subject to potential skin or eye contact with liquid AN.

- ❑ Training shall be provided at the time of initial assignment, or upon institution of the training program, and at least annually thereafter, and the employer shall assure that each employee is informed of the following:
  - The information contained in 29 CFR 1910.1045 and Appendixes A and B;
  - The quantity, location, manner of use, release, or storage of AN, and the specific nature of operations which could result in exposure to AN, as well as any necessary protective steps;
  - The purpose, proper use, and limitations of respirators and protective clothing;
  - The purpose and a description of the medical surveillance program;
  - The emergency procedures developed; and
  - Engineering and work-practice controls, their function, and the employee's relationship to these controls.

- ❑ The employer shall make a copy of 29 CFR 1910.1045 and its appendixes readily available to all affected employees.

- ❑ The employer shall provide, upon request, all materials relating to the employee information and training program to the Assistant Secretary and the Director.

- ❑ The employer may use labels or signs required by statutes, regulations, or ordinances in addition to, or in combination with, signs and labels.

- ❑ The employer shall assure that no statement appears on or near any sign or label which contradicts or detracts from the required sign or label.

- ❑ The employer shall post signs to clearly indicate all workplaces where AN concentrations exceed the permissible exposure limits. The signs should look like the sample at right.

- ❑ The employer shall assure that signs are illuminated and cleaned as necessary so that the legend is readily visible.

- ❑ The employer shall assure that precautionary labels are affixed to all containers of liquid AN and AN-based materials not exempted. The employer shall assure that the labels remain affixed when the materials are sold, distributed, or otherwise leave the employer's workplace.

- ❑ The employer shall assure that the precautionary labels are readily visible and legible.

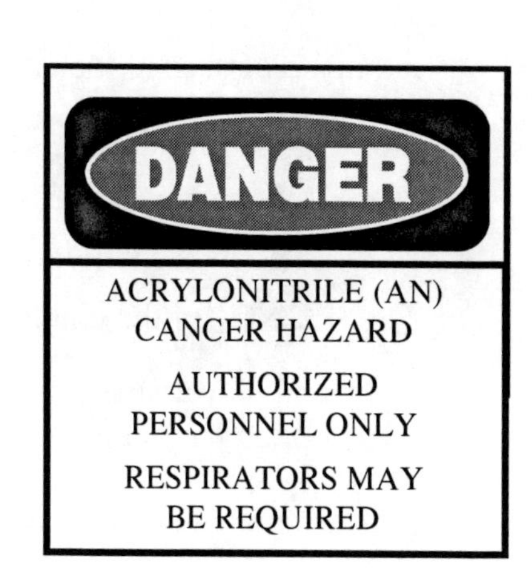

## Recordkeeping

- ❑ Where the processing, use, and handling of materials made from or containing AN are exempted, the employer shall establish and maintain an accurate record of objective data reasonably relied upon to support the exemption.
- ❑ This record shall include at least the following information:
  - The material qualifying for exemption;
  - The source of the objective data;
  - The testing protocol, results of testing, and/or analysis of the material for the release of AN;
  - A description of the operation exempted and how the data supports the exemption; and
  - Other data relevant to the operations, materials, and processing covered by the exemption.
- ❑ The employer shall maintain this record for the duration of the employer's reliance upon such objective data.
- ❑ The employer shall establish and maintain an accurate record of all monitoring.
- ❑ This record shall include:
  - The dates, number, duration, and results of each of the samples taken including a description of the sampling procedure used to determine representative employee exposure;
  - A description of the sampling and analytical methods used, and the data relied upon to establish that the methods used meet the accuracy and precision requirements;
  - Type of respiratory protective devices worn, if any; and
  - Name, social security number, and job classification of the employee monitored, and all other employees whose exposure the measurement is intended to represent.
- ❑ The employer shall maintain this record for at least 40 years, or for the duration of employment plus 20 years, whichever is longer.
- ❑ The employer shall establish and maintain an accurate record for each employee subject to medical surveillance.
- ❑ This record shall include:
  - A copy of the physician's written opinions;
  - Any employee medical complaints related to exposure to AN;
  - A copy of the information provided to the physician; and
  - A copy of the employee's medical and work history.
- ❑ The employer shall assure that this record be maintained for at least 40 years, or for the duration of employment plus 20 years, whichever is longer.
- ❑ The employer shall make all records required to be maintained by this section available, upon

request, to the Assistant Secretary and the Director for examination and copying.

- Records shall be provided upon request to employees, designated representatives, and the Assistant Secretary in accordance with 29 CFR 1910.20 (a) through (e) and (q) through (i). Records shall be provided in the same manner as exposure monitoring records.

- Whenever the employer ceases to do business, the successor employer shall receive and retain all records required to be maintained by this section for the prescribed period.

- Whenever the employer ceases to do business and there is no successor employer to receive and retain the records for the prescribed period, these records shall be transmitted to the Director.

- At the expiration of the retention period for the records required to be maintained pursuant to this section, the employer shall notify the Director at least three months prior to the disposal of the records, and shall transmit them to the Director upon request.

- The employer shall also comply with any additional requirements involving transfer of records set forth in 29 CFR 1910.20(h).

## Observation of Monitoring

- The employer shall provide affected employees, or their designated representatives, an opportunity to observe any monitoring of employee exposure to AN.

- Whenever observation of the monitoring of employee exposure to AN requires entry into an area where the use of protective clothing or equipment is required, the employer shall provide the observer with personal protective clothing and equipment required to be worn by employees working in the area, assure the use of such clothing and equipment, and require the observer to comply with all other applicable safety and health procedures.

- Without interfering with the monitoring, observers shall be entitled:
  - To receive an explanation of the measurement procedures;
  - To observe all steps related to the measurement of airborne concentrations of AN performed at the place of exposure; and
  - To record the results obtained.

# Air Contaminants

## 1910.1000

Section 4 provides a general overview regarding the management of airborne contaminants in the workplace. It provides the framework for understanding the subsequent sections that address specific contaminants on an individual basis. Also contained within this section are the Z-Tables for air contaminants. We recommend that the reader consult an industrial hygienist or other persons with the requisite knowledge and experience to perform air sampling.

### General

- Employers must limit any employee's exposure to substances listed in the 'Z' Tables, Tables 4-4-1 through 4-4-3.
- Exposure to substances with "C"-Ceiling Values should not at any time exceed the exposure limit given for that substance. When instantaneous monitoring is not possible, the ceiling shall be considered to be a 15-minute time-weighted average (TWA) exposure that cannot be exceeded at any time during the working day.
- Exposure to any substance in Table 4-4-1, that is not preceded by a "C", shall not exceed the 8-hour TWA given for that substance in any 8-hour work shift of a 40-hour work week.
- An employee's exposure to any substance listed in Table 4-4-2, in any 8-hour work shift of a 40-hour work week, must not exceed the 8-hour TWA limit in Table 4-4-2.
- An employee's exposure to a substance listed in Table 4-4-2 must not exceed the acceptable ceiling concentration in the table at any time during an 8-hour shift. An exception exists for a time period, and up to a concentration not exceeding the maximum duration and concentration allowed in the column under "acceptable maximum peak above the acceptable ceiling concentration for an 8-hour shift."

- An employee's exposure to any substance listed in Table 4-4-3, in any 8-hour work shift of a 40-hour workweek, must not exceed the 8-hour TWA limit in the table.

## Table 4-4-1: "Z-1" Limits for Air Contaminants

| Substance | CAS No. (c) | ppm (a)1 | mg/m³ (b)1 | Skin designation |
|---|---|---|---|---|
| Acetaldehyde | 75 - 07 - 0 | 200 | 360 | |
| Acetic acid | 64 - 19 - 7 | 10 | 25 | |
| Acetic anhydride | 108 - 24 - 7 | 5 | 20 | |
| Acetone | 67 - 64 - 1 | 1000 | 2400 | |
| Acetonitrile | 75 - 05 - 8 | 40 | 70 | |
| 2-Acetylaminofluorine; see 1910.1014 | 53 - 96 - 3 | | | |
| Acetylene dichloride; see 1,2-Dichloroethylene | | | | |
| Acetylene tetrabromide | 79 - 27 - 6 | 1 | 14 | |
| Acrolein | 107 - 02 - 8 | 0.1 | 0.25 | |
| Acrylamide | 79 - 06 - 1 | | 0.3 | X |
| Acrylonitrile; see 1910.1045 | 107 - 13 - 1 | | | |
| Aldrin | 309 - 00 - 2 | | 0.25 | X |
| Allyl alcohol | 107 - 18 - 6 | 2 | 5 | X |
| Allyl chloride | 107 - 05 - 1 | 1 | 3 | |
| Allyl glycidyl ether (AGE) | 106 - 92 - 3 | (C)10 | (C)45 | |
| Allyl propyl disulfide | 2179 - 59 - 1 | 2 | 12 | |
| alpha-Alumina | 1344 - 28 - 1 | | | |
| Total dust | | | 15 | |
| Respirable fraction | | | 5 | |
| Aluminum, metal (as Al) | 7429 - 90 - 5 | | | |
| Total dust | | | 15 | |
| Respirable fraction | | | 5 | |
| 4-Aminodiphenyl; see 1910.1011 | 92 - 67 - 1 | | | |
| 2-Aminoethanol; see Ethanolamine | | | | |
| 2-Aminopyridine | 504 - 29 - 0 | 0.5 | 2 | |
| Ammonia | 7664 - 41 - 7 | 50 | 35 | |
| Ammonium sulfamate | 7773 - 06 - 0 | | | |
| Total dust | | | 15 | |
| Respirable fraction | | | 5 | |
| n-Amyl acetate | 628 - 63 - 7 | 100 | 525 | |
| sec-Amyl acetate | 626 - 38 - 0 | 125 | 650 | |
| Aniline and homologs | 62 - 53 - 3 | 5 | 19 | X |
| Anisidine (o-, p-isomers) | 29191 - 52 - 4 | | 0.5 | X |
| Antimony and compounds (as Sb) | 7440 - 36 - 0 | | 0.5 | |
| ANTU (alpha Naphthylthiourea) | 86 - 88 - 4 | | 0.3 | |
| Arsenic, inorganic compounds (as As); see 1910.1018 | 7440 - 38 - 2 | | | |
| Arsenic, organic compounds (as As) | 7440 - 38 - 2 | | 0.5 | |
| Arsine | 7784 - 42 - 1 | 0.05 | 0.2 | |
| Asbestos; see 1910.1001 | (4) | | | |
| Azinphos-methyl | 86 - 50 - 0 | | 0.2 | X |
| Barium, soluble compounds (as Ba) | 7440 - 39 - 3 | | 0.5 | |
| Barium sulfate | 7727 - 43 - 7 | | | |

| Substance | CAS No. (c) | ppm (a)1 | mg/m³ (b)1 | Skin designation |
|---|---|---|---|---|
| Total dust | | | 15 | |
| Respirable fraction | | | 5 | |
| Benomyl | 17804 - 35 - 2 | | | |
| Total dust | | | 15 | |
| Respirable fraction | | | 5 | |
| Benzene; see 1910.1028 | 71 - 43 - 2 | | | |
| See Table Z-2 for the limits applicable in the operations or sectors excluded in 1910.1028[d] | | | | |
| Benzidine; see 1910.1010 | 92 - 87 - 5 | | | |
| p-Benzoquinone; see Quinone | | | | |
| Benzo(a)pyrene; see Coal tar pitch volatiles | | | | |
| Benzoyl peroxide | 94 - 36 - 0 | | 5 | |
| Benzyl chloride | 100 - 44 - 7 | 1 | 5 | |
| Beryllium and beryllium compounds (as Be) | 7440 - 41 - 7 | | (2) | |
| Biphenyl; see Diphenyl | | | | |
| Bismuth telluride, Undoped | 1304 - 82 - 1 | | | |
| Total dust | | | 15 | |
| Respirable fraction | | | 5 | |
| Boron oxide | 1303 - 86 - 2 | | | |
| Total dust | | | 15 | |
| Boron trifluoride | 7637 - 07 - 2 | (C)1 | (C)3 | |
| Bromine | 7726 - 95 - 6 | 0.1 | 0.7 | |
| Bromoform | 75 - 25 - 2 | 0.5 | 5 | X |
| Butadiene (1,3-Butadiene); see 29 CFR 1910.1051; 29 CFR 1910.19(l) | 106 - 99 - 0 | 1 ppm/5 ppm STEL | | |
| Butanethiol; see Butyl mercaptan | | | | |
| 2-Butanone (Methyl ethyl ketone) | 78 - 93 - 3 | 200 | 590 | |
| 2-Butoxyethanol | 111 - 76 - 2 | 50 | 240 | X |
| n-Butyl-acetate | 123 - 86 - 4 | 150 | 710 | |
| sec-Butyl acetate | 105 - 46 - 4 | 200 | 950 | |
| tert-Butyl acetate | 540 - 88 - 5 | 200 | 950 | |
| n-Butyl alcohol | 71 - 36 - 3 | 100 | 300 | |
| sec-Butyl alcohol | 78 - 92 - 2 | 150 | 450 | |
| tert-Butyl alcohol | 75 - 65 - 0 | 100 | 300 | |
| Butylamine | 109 - 73 - 9 | (C)5 | (C)15 | X |
| tert-Butyl chromate (as $CrO_3$) | 1189 - 85 - 1 | | (C)0.1 | X |
| n-Butyl glycidyl ether (BGE) | 2426 - 08 - 6 | 50 | 270 | |
| Butyl mercaptan | 109 - 79 - 5 | 10 | 35 | |
| p-tert-Butyltoluene | 98 - 51 - 1 | 10 | 60 | |
| Cadmium (as Cd); see 1910.1027 | 7440 - 43 - 9 | | | |
| Calcium carbonate | 1317 - 65 - 3 | | | |
| Total dust | | | 15 | |
| Respirable fraction | | | 5 | |
| Calcium hydroxide | 1305 - 62 - 0 | | | |
| Total dust | | | 15 | |
| Respirable fraction | | | 5 | |
| Calcium oxide | 1305 - 78 - 8 | | 5 | |
| Calcium silicate | 1344 - 95 - 2 | | | |

| Substance | CAS No. (c) | ppm (a)1 | mg/m3 (b)1 | Skin designation |
|---|---|---|---|---|
| Total dust | | | 15 | |
| Respirable fraction | | | 5 | |
| Calcium sulfate | 7778 - 18 - 9 | | | |
| Total dust | | | 15 | |
| Respirable fraction | | | 5 | |
| Camphor, synthetic | 76 - 22 - 2 | | 2 | |
| Carbaryl (Sevin) | 63 - 25 - 2 | | 5 | |
| Carbon black | 1333 - 86 - 4 | | 3.5 | |
| Carbon dioxide | 124 - 38 - 9 | 5000 | 9000 | |
| Carbon disulfide | 75 - 15 - 0 | | (2) | |
| Carbon monoxide | 630 - 08 - 0 | 50 | 55 | |
| Carbon tetrachloride | 56 - 23 - 5 | | (2) | |
| Cellulose | 9004 - 34 - 6 | | | |
| Total dust | | | 15 | |
| Respirable fraction | | | 5 | |
| Chlordane | 57 - 74 - 9 | | 0.5 | X |
| Chlorinated camphene | 8001 - 35 - 2 | | 0.5 | X |
| Chlorinated diphenyl-oxide | 55720 - 99 - 5 | | 0.5 | |
| Chlorine | 7782 - 50 - 5 | (C)1 | (C)3 | |
| Chlorine dioxide | 10049 - 04 - 4 | 0.1 | 0.3 | |
| Chlorine trifluoride | 7790 - 91 - 2 | (C)0.1 | (C)0.4 | |
| Chloroacetaldehyde | 107 - 20 - 0 | (C)1 | (C)3 | |
| a-Chloroacetophenone (Phenacyl chloride) | 532 - 27 - 4 | 0.05 | 0.3 | |
| Chlorobenzene | 108 - 90 - 7 | 75 | 350 | |
| o-Chlorobenzylidene-malononitrile | 2698 - 41 - 1 | 0.05 | 0.4 | |
| Chlorobromomethane | 74 - 97 - 5 | 200 | 1050 | |
| 2-Chloro-1,3-butadiene; see beta-Chloroprene | | | | |
| Chlorodiphenyl (42%Chlorine) (PCB) | 53469 - 21 - 9 | | 1 | X |
| Chlorodiphenyl (54%Chlorine) (PCB) | 11097 - 69 - 1 | | 0.5 | X |
| 1-Chloro-2,3-epoxypropane; see Epichlorohydrin | | | | |
| 2-Chloroethanol; see Ethylene chlorohydrin | | | | |
| Chloroethylene; see Vinyl chloride | | | | |
| Chloroform (Trichloromethane) | 67 - 66 - 3 | (C)50 | (C)240 | |
| bis(Chloromethyl) ether; see 1910.1008 | 542 - 88 - 1 | | | |
| Chloromethyl methyl ether; see 1910.1006 | 107 - 30 - 2 | | | |
| 1-Chloro-1-nitropropane | 600 - 25 - 9 | 20 | 100 | |
| Chloropicrin | 76 - 06 - 2 | 0.1 | 0.7 | |
| beta-Chloroprene | 126 - 99 - 8 | 25 | 90 | X |
| 2-Chloro-6-(trichloromethyl) pyridine | 1929 - 82 - 4 | | | |
| Total dust | | | 15 | |
| Respirable fraction | | | 5 | |
| Chromic acid and chromates (as $CrO_3$) | (4) | | (2) | |
| Chromium (II) compounds (as Cr) | 7440 - 47 - 3 | | 0.5 | |
| Chromium (III) compounds (as Cr) | 7440 - 47 - 3 | | 0.5 | |
| Chromium metal and insol. salts (as Cr) | 7440 - 47 - 3 | | 1 | |
| Chrysene; see Coal tar pitch volatiles | | | | |
| Clopidol | 2971 - 90 - 6 | | | |

| Substance | CAS No. (c) | ppm (a)1 | mg/m3 (b)1 | Skin designation |
|---|---|---|---|---|
| Total dust | | | 15 | |
| Respirable fraction | | | 5 | |
| Coal dust (less than 5% SiO2), respirable fraction | | | (3) | |
| Coal dust (greater than or equal to 5% SiO2), respirable fraction | | | (3) | |
| Coal tar pitch volatiles (benzene soluble fraction), anthracene, BaP, phenanthrene, acridine, chrysene, pyrene | 65966 - 93 - 2 | | 0.2 | |
| Cobalt metal, dust, and fume (as Co) | 7440 - 48 - 4 | | 0.1 | |
| Coke oven emissions; see 1910.1029 | | | | |
| Copper | 7440 - 50 - 8 | | | |
| Fume (as Cu) | | | 0.1 | |
| Dusts and mists (as Cu) | | | 1 | |
| Cotton duste; see 1910.1043 | | | 1 | |
| Crag herbicide (Sesone) | 136 - 78 - 7 | | | |
| Total dust | | | 15 | |
| Respirable fraction | | | 5 | |
| Cresol, all isomers | 1319 - 77 - 3 | 5 | 22 | X |
| Crotonaldehyde | 123 - 73 - 9; 4170 - 30 - 3 | 2 | 6 | |
| Cumene | 98 - 82 - 8 | 50 | 245 | X |
| Cyanides (as CN) | (4) | | 5 | X |
| Cyclohexane | 110 - 82 - 7 | 300 | 1050 | |
| Cyclohexanol | 108 - 93 - 0 | 50 | 200 | |
| Cyclohexanone | 108 - 94 - 1 | 50 | 200 | |
| Cyclohexene | 110 - 83 - 8 | 300 | 1015 | |
| Cyclopentadiene | 542 - 92 - 7 | 75 | 200 | |
| 2,4-D (Dichlorophenoxyacetic acid) | 94 - 75 - 7 | | 10 | |
| Decaborane | 17702 - 41 - 9 | 0.05 | 0.3 | X |
| Demeton (Systox) | 8065 - 48 - 3 | | 0.1 | X |
| Diacetone alcohol (4-Hydroxy-4-methyl-2-pentanone) | 123 - 42 - 2 | 50 | 240 | |
| 1,2-Diaminoethane; see Ethylenediamine | | | | |
| Diazomethane | 334 - 88 - 3 | 0.2 | 0.4 | |
| Diborane | 19287 - 45 - 7 | 0.1 | 0.1 | |
| 1,2-Dibromo-3- chloropropane (DBCP); see 1910.1044 | 96 - 12 - 8 | | | |
| 1,2-Dibromoethane; see Ethylene dibromide | | | | |
| Dibutyl phosphate | 107 - 66 - 4 | 1 | 5 | |
| Dibutyl phthalate | 84 - 74 - 2 | | 5 | |
| o-Dichlorobenzene | 95 - 50 - 1 | (C)50 | (C)300 | |
| p-Dichlorobenzene | 106 - 46 - 7 | 75 | 450 | |
| 3,'-Dichlorobenzidine; see 1910.1007 | 91 - 94 - 1 | | | |
| Dichlorodifluoromethane | 75 - 71 - 8 | 1000 | 4950 | |
| 1,3-Dichloro-5,5-dimethyl hydantoin | 118 - 52 - 5 | | 0.2 | |
| Dichlorodiphenyltri-chloroethane (DDT) | 50 - 29 - 3 | | 1 | X |
| 1,1-Dichloroethane | 75 - 34 - 3 | 100 | 400 | |
| 1,2-Dichloroethane; see Ethylene dichloride | | | | |
| 1,2-Dichloroethylene | 540 - 59 - 0 | 200 | 790 | |
| Dichloroethyl ether | 111 - 44 - 4 | (C)15 | (C)90 | X |
| Dichloromethane; see Methylene chloride | | | | |
| Dichloromonofluoro-methane | 75 - 43 - 4 | 1000 | 4200 | |
| 1,1-Dichloro-1-nitroethane | 594 - 72 - 9 | (C)10 | (C)60 | |

| Substance | CAS No. (c) | ppm (a)1 | mg/m3 (b)1 | Skin designation |
|---|---|---|---|---|
| 1,2-Dichloropropane; see Propylene dichloride | | | | |
| Dichlorotetrafluoro-ethane | 76 - 14 - 2 | 1000 | 7000 | |
| Dichlorvos (DDVP) | 62 - 73 - 7 | | 1 | X |
| Dicyclopentadienyl iron | 102 - 54 - 5 | | | |
| Total dust | 15 | | | |
| Respirable fraction | 5 | | | |
| Dieldrin | 60 - 57 - 1 | | 0.25 | X |
| Diethylamine | 109 - 89 - 7 | 25 | 75 | |
| 2-Diethylaminoethanol | 100 - 37 - 8 | 10 | 50 | X |
| Diethyl ether; see Ethyl ether | | | | |
| Difluorodibromomethane | 75 - 61 - 6 | 100 | 860 | |
| Diglycidyl ether (DGE) | 2238 - 07 - 5 | (C)0.5 | (C)2.8 | |
| Dihydroxybenzene; see Hydroquinone | | | | |
| Diisobutyl ketone | 108 - 83 - 8 | 50 | 290 | |
| Diisopropylamine | 108 - 18 - 9 | 5 | 20 | X |
| 4-Dimethylaminoazobenzene; see 1910.1015 | 60 - 11 - 7 | | | |
| Dimethoxymethane; see Methylal | | | | |
| Dimethyl acetamide | 127 - 19 - 5 | 10 | 35 | X |
| Dimethylamine | 124 - 40 - 3 | 10 | 18 | |
| Dimethylaminobenzene; see Xylidine | | | | |
| Dimethylaniline (N,N-Dimethylaniline) | 121 - 69 - 7 | 5 | 25 | X |
| Dimethylbenzene; see Xylene | | | | |
| Dimethyl-1,2-dibromo-2,2-dichloroethyl phosphate | 300 - 76 - 5 | | 3 | |
| Dimethylformamide | 68 - 12 - 2 | 10 | 30 | X |
| 2,6-Dimethyl-4-heptanone; see Diisobutyl ketone | | | | |
| 1,1-Dimethylhydrazine | 57 - 14 - 7 | 0.5 | 1 | X |
| Dimethylphthalate | 131 - 11 - 3 | | 5 | |
| Dimethyl sulfate | 77 - 78 - 1 | 1 | 5 | X |
| Dinitrobenzene (all isomers) | | | 1 | X |
| (ortho) | 528 - 29 - 0 | | | |
| (meta) | 99 - 65 - 0 | | | |
| (para) | 100 - 25 - 4 | | | |
| Dinitro-o-cresol | 534 - 52 - 1 | | 0.2 | X |
| Dinitrotoluene | 25321 - 14 - 6 | | 1.5 | X |
| Dioxane (Diethylene dioxide) | 123 - 91 - 1 | 100 | 360 | X |
| Diphenyl (Biphenyl) | 92 - 52 - 4 | 0.2 | 1 | |
| Diphenylmethane | | | | |
| diisocyanate; see Methylene bisphenyl isocyanate | | | | |
| Dipropylene glycol methyl ether | 34590 - 94 - 8 | 100 | 600 | X |
| Di-sec octyl phthalate (Di-(2-ethylhexyl) phthalate) | 117 - 81 - 7 | | 5 | |
| Emery | 12415 - 34 - 8 | | | |
| Total dust | | | 15 | |
| Respirable fraction | | | 5 | |
| Endrin | 72 - 20 - 8 | | 0.1 | X |
| Epichlorohydrin | 106 - 89 - 8 | 5 | 19 | X |
| EPN | 2104 - 64 - 5 | | 0.5 | X |
| 1,2-Epoxypropane; see Propylene oxide | | | | |
| 2,3-Epoxy-1-propanol; see Glycidol | | | | |
| Ethanethiol; see Ethyl mercaptan | | | | |
| Ethanolamine | 141 - 43 - 5 | 3 | 6 | |

| Substance | CAS No. (c) | ppm (a)1 | mg/m3 (b)1 | Skin designation |
|---|---|---|---|---|
| 2-Ethoxyethanol (Cellosolve) | 110 - 80 - 5 | 200 | 740 | X |
| 2-Ethoxyethyl acetate (Cellosolve acetate) | 111 - 15 - 9 | 100 | 540 | X |
| Ethyl acetate | 141 - 78 - 6 | 400 | 1400 | |
| Ethyl acrylate | 140 - 88 - 5 | 25 | 100 | X |
| Ethyl alcohol (Ethanol) | 64 - 17 - 5 | 1000 | 1900 | |
| Ethylamine | 75 - 04 - 7 | 10 | 18 | |
| Ethyl amyl ketone (5-Methyl-3-heptanone) | 541 - 85 - 5 | 25 | 130 | |
| Ethyl benzene | 100 - 41 - 4 | 100 | 435 | |
| Ethyl bromide | 74 - 96 - 4 | 200 | 890 | |
| Ethyl butyl ketone (3-Heptanone) | 106 - 35 - 4 | 50 | 230 | |
| Ethyl chloride | 75 - 00 - 3 | 1000 | 2600 | |
| Ethyl ether | 60 - 29 - 7 | 400 | 1200 | |
| Ethyl formate | 109 - 94 - 4 | 100 | 300 | |
| Ethyl mercaptan | 75 - 08 - 1 | (C)10 | (C)25 | |
| Ethyl silicate | 78 - 10 - 4 | 100 | 850 | |
| Ethylene chlorohydrin | 107 - 07 - 3 | 5 | 16 | X |
| Ethylenediamine | 107 - 15 - 3 | 10 | 25 | |
| Ethylene dibromide | 106 - 93 - 4 | | (2) | |
| Ethylene dichloride (1,2-Dichloroethane) | 107 - 06 - 2 | | (2) | |
| Ethylene glycol dinitrate | 628 - 96 - 6 | (C)0.2 | (C)1 | X |
| Ethylene glycol methyl acetate; see Methyl cellosolve acetate | | | | |
| Ethyleneimine; see 1910.1012 | 151 - 56 - 4 | | | |
| Ethylene oxide; see 1910.1047 | 75 - 21 - 8 | | | |
| Ethylidene chloride; see 1,1-Dichloroethane | | | | |
| N-Ethylmorpholine | 100 - 74 - 3 | 20 | 94 | X |
| Ferbam | 14484 - 64 - 1 | | | |
| Total dust | 15 | | | |
| Ferrovanadium dust | 12604 - 58 - 9 | | 1 | |
| Fluorides (as F) | (4) | | 2.5 | |
| Fluorine | 7782 - 41 - 4 | 0.1 | 0.2 | |
| Fluorotrichloromethane (Trichlorofluoromethane) | 75 - 69 - 4 | 1000 | 5600 | |
| Formaldehyde; see 1910.1048 | 50 - 00 - 0 | | | |
| Formic acid | 64 - 18 - 6 | 5 | 9 | |
| Furfural | 98 - 01 - 1 | 5 | 20 | X |
| Furfuryl alcohol | 98 - 00 - 0 | 50 | 200 | |
| Grain dust (oat, wheat, barley) | | | 10 | |
| Glycerin (mist) | 56 - 81 - 5 | | | |
| Total dust | | | 15 | |
| Respirable fraction | | | 5 | |
| Glycidol | 556 - 52 - 5 | 50 | 150 | |
| Glycol monoethyl ether; see 2-Ethoxyethanol | | | | |
| Graphite, natural, respirable dust | 7782 - 42 - 5 | | (3) | |
| Graphite, synthetic | | | | |
| Total dust | | | 15 | |
| Respirable fraction | | | 5 | |
| Guthion; see Azinphos methyl | | | | |
| Gypsum | 13397 - 24 - 5 | | | |

| Substance | CAS No. (c) | ppm (a)1 | mg/m3 (b)1 | Skin designation |
|---|---|---|---|---|
| Total dust | | | 15 | |
| Respirable fraction | | | 5 | |
| Hafnium | 7440 - 58 - 6 | | 0.5 | |
| Heptachlor | 76 - 44 - 8 | | 0.5 | X |
| Heptane (n-Heptane) | 142 - 82 - 5 | 500 | 2000 | |
| Hexachloroethane | 67 - 72 - 1 | 1 | 10 | X |
| Hexachloronaphthalene | 1335 - 87 - 1 | | 0.2 | X |
| n-Hexane | 110 - 54 - 3 | 500 | 1800 | |
| 2-Hexanone (Methyl n-butyl ketone) | 591 - 78 - 6 | 100 | 410 | |
| Hexone (Methyl isobutyl ketone) | 108 - 10 - 1 | 100 | 410 | |
| sec-Hexyl acetate | 108 - 84 - 9 | 50 | 300 | |
| Hydrazine | 302 - 01 - 2 | 1 | 1.3 | X |
| Hydrogen bromide | 10035 - 10 - 6 | 3 | 10 | |
| Hydrogen chloride | 7647 - 01 - 0 | (C)5 | (C)7 | |
| Hydrogen cyanide | 74 - 90 - 8 | 10 | 11 | X |
| Hydrogen fluoride (as F) | 7664 - 39 - 3 | | (2) | |
| Hydrogen peroxide | 7722 - 84 - 1 | 1 | 1.4 | |
| Hydrogen selenide (as Se) | 7783 - 07 - 5 | 0.05 | 0.2 | |
| Hydrogen sulfide | 7783 - 06 - 4 | | (2) | |
| Hydroquinone | 123 - 31 - 9 | | 2 | |
| Iodine | 7553 - 56 - 2 | (C)0.1 | (C)1 | |
| Iron oxide fume | 1309 - 37 - 1 | | 10 | |
| Isoamyl acetate | 123 - 92 - 2 | 100 | 525 | |
| Isoamyl alcohol (primary and secondary) | 123 - 51 - 3 | 100 | 360 | |
| Isobutyl acetate | 110 - 19 - 0 | 150 | 700 | |
| Isobutyl alcohol | 78 - 83 - 1 | 100 | 300 | |
| Isophorone | 78 - 59 - 1 | 25 | 140 | |
| Isopropyl acetate | 108 - 21 - 4 | 250 | 950 | |
| Isopropyl alcohol | 67 - 63 - 0 | 400 | 980 | |
| Isopropylamine | 75 - 31 - 0 | 5 | 12 | |
| Isopropyl ether | 108 - 20 - 3 | 500 | 2100 | |
| Isopropyl glycidyl ether (IGE) | 4016 - 14 - 2 | 50 | 240 | |
| Kaolin | 1332 - 58 - 7 | | | |
| Total dust | | | 15 | |
| Respirable fraction | | | 5 | |
| Ketene | 463 - 51 - 4 | 0.5 | 0.9 | |
| Lead, inorganic (as Pb); see 1910.1025 | 7439 - 92 - 1 | | | |
| Limestone | 1317 - 65 - 3 | | | |
| Total dust | | | 15 | |
| Respirable fraction | | | 5 | |
| Lindane | 58 - 89 - 9 | | 0.5 | X |
| Lithium hydride | 7580 - 67 - 8 | | 0.025 | |
| L.P.G. (Liquefied petroleum gas) | 68476 - 85 - 7 | 1000 | 1800 | |
| Magnesite | 546 - 93 - 0 | | | |
| Total dust | | | 15 | |
| Respirable fraction | | | 5 | |
| Magnesium oxide fume | 1309 - 48 - 4 | | | |
| Total particulate | | | 15 | |
| Malathion | 121 - 75 - 5 | | | |

| Substance | CAS No. (c) | ppm (a)1 | mg/m3 (b)1 | Skin designation |
|---|---|---|---|---|
| Total dust | | 15 | X | |
| Maleic anhydride | 108 - 31 - 6 | 0.25 | 1 | |
| Manganese compounds (as Mn) | 7439 - 96 - 5 | (C)5 | | |
| Manganese fume (as Mn) | 7439 - 96 - 5 | (C)5 | | |
| Marble | 1317 - 65 - 3 | | | |
| Total dust | | | 15 | |
| Respirable fraction | | | 5 | |
| Mercury (aryl and inorganic) (as Hg) | 7439 - 97 - 6 | | (2) | |
| Mercury (organo) alkyl compounds (as Hg) | 7439 - 97 - 6 | | (2) | |
| Mercury (vapor) (as Hg) | 7439 - 97 - 6 | | (2) | |
| Mesityl oxide | 141 - 79 - 7 | 25 | 100 | |
| Methanethiol; see Methyl mercaptan | | | | |
| Methoxychlor | 72 - 43 - 5 | | | |
| Total dust | | | 15 | |
| 2-Methoxyethanol (Methyl cellosolve) | 109 - 86 - 4 | 25 | 80 | X |
| 2-Methoxyethyl acetate (Methyl cellosolve acetate) | 110 - 49 - 6 | 25 | 120 | X |
| Methyl acetate | 79 - 20 - 9 | 200 | 610 | |
| Methyl acetylene (Propyne) | 74 - 99 - 7 | 1000 | 1650 | |
| Methyl acetylene-propadiene mixture (MAPP) | | 1000 | 1800 | |
| Methyl acrylate | 96 - 33 - 3 | 10 | 35 | X |
| Methylal (Dimethoxy-methane) | 109 - 87 - 5 | 1000 | 3100 | |
| Methyl alcohol | 67 - 56 - 1 | 200 | 260 | |
| Methylamine | 74 - 89 - 5 | 10 | 12 | |
| Methyl amyl alcohol; see Methyl isobutyl carbinol | | | | |
| Methyl n-amyl ketone | 110 - 43 - 0 | 100 | 465 | |
| Methyl bromide | 74 - 83 - 9 | (C)20 | (C)80 | X |
| Methyl butyl ketone; see 2-Hexanone | | | | |
| Methyl cellosolve; see 2-Methoxyethanol | | | | |
| Methyl cellosolve acetate; see 2-Methoxyethyl acetate | | | | |
| Methyl chloride | 74 - 87 - 3 | | (2) | |
| Methyl chloroform (1,1,1-Trichloroethane) | 71 - 55 - 6 | 350 | 1900 | |
| Methylcyclohexane | 108 - 87 - 2 | 500 | 2000 | |
| Methylcyclohexanol | 25639 - 42 - 3 | 100 | 470 | |
| o-Methylcyclohexanone | 583 - 60 - 8 | 100 | 460 | X |
| Methylene chloride | 75 - 09 - 2 | | (2) | |
| Methyl ethyl ketone (MEK); see 2-Butanone | | | | |
| Methyl formate | 107 - 31 - 3 | 100 | 250 | |
| Methyl hydrazine (Monomethyl hydrazine) | 60 - 34 - 4 | (C)0.2 | (C)0.35 | X |
| Methyl iodide | 74 - 88 - 4 | 5 | 28 | X |
| Methyl isoamyl ketone | 110 - 12 - 3 | 100 | 475 | |
| Methyl isobutyl carbinol | 108 - 11 - 2 | 25 | 100 | X |
| Methyl isobutyl ketone; see Hexone | | | | |
| Methyl isocyanate | 624 - 83 - 9 | 0.02 | 0.05 | X |
| Methyl mercaptan | 74 - 93 - 1 | (C)10 | (C)20 | |
| Methyl methacrylate | 80 - 62 - 6 | 100 | 410 | |
| Methyl propyl ketone; see 2-Pentanone | | | | |
| alpha-Methyl styrene | 98 - 83 - 9 | (C)100 | (C)480 | |

| Substance | CAS No. (c) | ppm (a)1 | mg/m3 (b)1 | Skin designation |
|---|---|---|---|---|
| Methylene bisphenyl isocyanate (MDI) | 101 - 68 - 8 | (C)0.02 | (C)0.2 | |
| Mica; see Silicates. | | | | |
| Molybdenum (as Mo) | 7439 - 98 - 7 | | | |
| Soluble compounds | 5 | | | |
| Insoluble compounds | | | | |
| Total dust | 15 | | | |
| Monomethyl aniline | 100 - 61 - 8 | 2 | 9 | X |
| Monomethyl hydrazine; see Methyl hydrazine | | | | |
| Morpholine | 110 - 91 - 8 | 20 | 70 | X |
| Naphtha (Coal tar) | 8030 - 30 - 6 | 100 | 400 | |
| Naphthalene | 91 - 20 - 3 | 10 | 50 | |
| alpha-Naphthylamine; see 1910.1004 | 134 - 32 - 7 | | | |
| beta-Naphthylamine; see 1910.1009 | 91 - 59 - 8 | | | |
| Nickel carbonyl (as Ni) | 13463 - 39 - 3 | 0.001 | 0.007 | |
| Nickel, metal and insoluble compounds (as Ni) | 7440 - 02 - 0 | 1 | | |
| Nickel, soluble compounds (as Ni) | 7440 - 02 - 0 | 1 | | |
| Nicotine | 54 - 11 - 5 | 0.5 | X | |
| Nitric acid | 7697 - 37 - 2 | 2 | 5 | |
| Nitric oxide | 10102 - 43 - 9 | 25 | 30 | |
| p-Nitroaniline | 100 - 01 - 6 | 1 | 6 | X |
| Nitrobenzene | 98 - 95 - 3 | 1 | 5 | X |
| p-Nitrochlorobenzene | 100 - 00 - 5 | 1 | X | |
| 4-Nitrodiphenyl; see 1910.1003 | 92 - 93 - 3 | | | |
| Nitroethane | 79 - 24 - 3 | 100 | 310 | |
| Nitrogen dioxide | 10102 - 44 - 0 | (C)5 | (C)9 | |
| Nitrogen trifluoride | 7783 - 54 - 2 | 10 | 29 | |
| Nitroglycerin | 55 - 63 - 0 | (C)0.2 | (C)2 | X |
| Nitromethane | 75 - 52 - 5 | 100 | 250 | |
| 1-Nitropropane | 108 - 03 - 2 | 25 | 90 | |
| 2-Nitropropane | 79 - 46 - 9 | 25 | 90 | |
| N-Nitrosodimethylamine ; see 1910.1016 | | | | |
| Nitrotoluene (all isomers) | | 5 | 30 | X |
| o-isomer | 88 - 72 - 2 | | | |
| m-isomer | 99 - 08 - 1 | | | |
| p-isomer | 99 - 99 - 0 | | | |
| Nitrotrichloromethane; see Chloropicrin | | | | |
| Octachloronaphthalene | 2234 - 13 - 1 | | 0.1 | X |
| Octane | 111 - 65 - 9 | 500 | 2350 | |
| Oil mist, mineral | 8012 - 95 - 1 | | 5 | |
| Osmium tetroxide (as Os) | 20816 - 12 - 0 | 0.002 | | |
| Oxalic acid | 144 - 62 - 7 | | 1 | |
| Oxygen difluoride | 7783 - 41 - 7 | 0.05 | 0.1 | |
| Ozone | 10028 - 15 - 6 | 0.1 | 0.2 | |
| Paraquat, respirable dust | 4685 - 14 - 7;<br>1910 - 42 - 5;<br>2074 - 50 - 2 | | 0.5 | X |
| Parathion | 56 - 38 - 2 | | 0.1 | X |
| Particulates not otherwise regulated (PNOR) | | | | |

| Substance | CAS No. (c) | ppm (a)1 | mg/m3 (b)1 | Skin designation |
|---|---|---|---|---|
| Total dust | | | 15 | |
| Respirable fraction | | | 5 | |
| PCB; see Chlorodiphenyl (42% and 54% chlorine) | | | | |
| Pentaborane | 19624 - 22 - 7 | 0.005 | 0.01 | |
| Pentachloronaphthalene | 1321 - 64 - 8 | | 0.5 | X |
| Pentachlorophenol | 87 - 86 - 5 | | 0.5 | X |
| Pentaerythritol | 115 - 77 - 5 | | | |
| Total dust | | | 15 | |
| Respirable fraction | | | 5 | |
| Pentane | 109 - 66 - 0 | 1000 | 2950 | |
| 2-Pentanone (Methyl propyl ketone) | 107 - 87 - 9 | 200 | 700 | |
| Perchloroethylene (Tetrachloroethylene) | 127 - 18 - 4 | | (2) | |
| Perchloromethyl mercaptan | 594 - 42 - 3 | 0.1 | 0.8 | |
| Perchloryl fluoride | 7616 - 94 - 6 | 3 | 13.5 | |
| Petroleum distillates (Naphtha) (Rubber Solvent) | | 500 | 2000 | |
| Phenol | 108 - 95 - 2 | 5 | 19 | X |
| p-Phenylene diamine | 106 - 50 - 3 | | 0.1 | X |
| Phenyl ether, vapor | 101 - 84 - 8 | 1 | 7 | |
| Phenyl ether-biphenyl mixture, vapor | | 1 | 7 | |
| Phenylethylene; see Styrene | | | | |
| Phenyl glycidyl ethe (PGE) | 122 - 60 - 1 | 10 | 60 | |
| Phenylhydrazine | 100 - 63 - 0 | 5 | 22 | X |
| Phosdrin (Mevinphos) | 7786 - 34 - 7 | | 0.1 | X |
| Phosgene (Carbonyl chloride) | 75 - 44 - 5 | 0.1 | 0.4 | |
| Phosphine | 7803 - 51 - 2 | 0.3 | 0.4 | |
| Phosphoric acid | 7664 - 38 - 2 | | 1 | |
| Phosphorus (yellow) | 7723 - 14 - 0 | | 0.1 | |
| Phosphorus pentachloride | 10026 - 13 - 8 | | 1 | |
| Phosphorus pentasulfide | 1314 - 80 - 3 | | 1 | |
| Phosphorus trichloride | 7719 - 12 - 2 | 0.5 | 3 | |
| Phthalic anhydride | 85 - 44 - 9 | 2 | 12 | |
| Picloram | 1918 - 02 - 1 | | | |
| Total dust | | | 15 | |
| Respirable fraction | | | 5 | |
| Picric acid | 88 - 89 - 1 | | 0.1 | X |
| Pindone (2-Pivalyl-1,3-indandione) | 83 - 26 - 1 | | 0.1 | |
| Plaster of Paris | 26499 - 65 - 0 | | | |
| Total dust | | | 15 | |
| Respirable fraction | | | 5 | |
| Platinum (as Pt) | 7440 - 06 - 4 | | | |
| Metal | | | | |
| Soluble salts | | | 0.002 | |
| Portland cement | 65997 - 15 - 1 | | | |
| Total dust | | | 15 | |
| Respirable fraction | | | 5 | |
| Propane | 74 - 98 - 6 | 1000 | 1800 | |
| beta-Propriolactone; see 1910.1013 | 57 - 57 - 8 | | | |
| n-Propyl acetate | 109 - 60 - 4 | 200 | 840 | |
| n-Propyl alcohol | 71 - 23 - 8 | 200 | 500 | |
| n-Propyl nitrate | 627 - 13 - 4 | 25 | 110 | |
| Propylene dichloride | 78 - 87 - 5 | 75 | 350 | |

| Substance | CAS No. (c) | ppm (a)1 | mg/m3 (b)1 | Skin designation |
|---|---|---|---|---|
| Propylene imine | 75 - 55 - 8 | 2 | 5 | X |
| Propylene oxide | 75 - 56 - 9 | 100 | 240 | |
| Propyne; see Methyl acetylene | | | | |
| Pyrethrum | 8003 - 34 - 7 | | 5 | |
| Pyridine | 110 - 86 - 1 | 5 | 15 | |
| Quinone | 106 - 51 - 4 | 0.1 | 0.4 | |
| RDX; see Cyclonite. | | | | |
| Rhodium (as Rh), metal fume and insoluble compounds | 7440 - 16 - 6 | | 0.1 | |
| Rhodium (as Rh), soluble compounds | 7440 - 16 - 6 | | 0.001 | |
| Ronnel | 299 - 84 - 3 | | 15 | |
| Rotenone | 83 - 79 - 4 | | 5 | |
| Rouge | | | | |
| Total dust | | | 15 | |
| Respirable fraction | | | 5 | |
| Selenium compounds (as Se) | 7782 - 49 - 2 | | 0.2 | |
| Selenium hexafluoride (as Se) | 7783 - 79 - 1 | 0.05 | 0.4 | |
| Silica, amorphous, precipitated and gel | 112926 - 00 - 8 | | (3) | |
| Silica, amorphous, diatomaceous earth, containing less than 1% crystalline silica | 61790 - 53 - 2 | | (3) | |
| Silica, crystalline cristobalite, respirable dust | 14464 - 46 - 1 | | (3) | |
| Silica, crystalline quartz, respirable dust | 14808 - 60 - 7 | | (3) | |
| Silica, crystalline tripoli (as quartz), respirable dust | 1317 - 95 - 9 | | (3) | |
| Silica, crystalline tridymite, respirable dust | 15468 - 32 - 3 | | (3) | |
| Silica, fused, respirable dust | 60676 - 86 - 0 | | (3) | |
| Silicates (less than 1% crystalline silica) | | | | |
| Mica (respirable dust) | 12001 - 26 - 2 | | (3) | |
| Soapstone, total dust | | | (3) | |
| Soapstone, respirable dust | | | (3) | |
| Talc (containing asbestos); use asbestos limit; see 29 CFR 1910.1001 | | | (3) | |
| Talc (containing no asbestos), respirable dust | 14807 - 96 - 6 | | (3) | |
| Tremolite, asbestiform; see 1910.1001 | | | | |
| Silicon | 7440 - 21 - 3 | | | |
| Total dust | | | 15 | |
| Respirable fraction | | | 5 | |
| Silicon carbide | 409 - 21 - 2 | | | |
| Total dust | | | 15 | |
| Respirable fraction | | | 5 | |
| Silver, metal and soluble compounds (as Ag) | 7440 - 22 - 4 | | 0.01 | |
| Soapstone; see Silicates | | | | |
| Sodium fluoroacetate | 62 - 74 - 8 | | 0.05 | X |
| Sodium hydroxide | 1310 - 73 - 2 | | 2 | |
| Starch | 9005 - 25 - 8 | | | |
| Total dust | | | 15 | |
| Respirable fraction | | | 5 | |
| Stibine | 7803 - 52 - 3 | 0.1 | 0.5 | |
| Stoddard solvent | 8052 - 41 - 3 | 500 | 2900 | |
| Strychnine | 57 - 24 - 9 | | 0.15 | |
| Styrene | 100 - 42 - 5 | | (2) | |

| Substance | CAS No. (c) | ppm (a)1 | mg/m3 (b)1 | Skin designation |
|---|---|---|---|---|
| Sucrose | 57 - 50 - 1 | | | |
| Total dust | | | 15 | |
| Respirable fraction | | | 5 | |
| Sulfur dioxide | 7446 - 09 - 5 | 5 | 13 | |
| Sulfur hexafluoride | 2551 - 62 - 4 | 1000 | 6000 | |
| Sulfuric acid | 7664 - 93 - 9 | | 1 | |
| Sulfur monochloride | 10025 - 67 - 9 | 1 | 6 | |
| Sulfur pentafluoride | 5714 - 22 - 7 | 0.025 | 0.25 | |
| Sulfuryl fluoride | 2699 - 79 - 8 | 5 | 20 | |
| Systox; see Demeton | | | | |
| 2,4,5-T (2,4,5-trichlorophenoxyacetic acid) | 93 - 76 - 5 | | 10 | |
| Talc; see Silicates | | | | |
| Tantalum, metal and oxide dust | 7440 - 25 - 7 | | 5 | |
| TEDP (Sulfotep) | 3689 - 24 - 5 | | 0.2 | X |
| Tellurium and compounds (as Te) | 13494 - 80 - 9 | | 0.1 | |
| Tellurium hexafluoride (as Te) | 7783 - 80 - 4 | 0.02 | 0.2 | |
| Temephos | 3383 - 96 - 8 | | | |
| Total dust | | | 15 | |
| Respirable fraction | | | 5 | |
| TEPP (Tetraethyl pyrophosphate) | 107 - 49 - 3 | | 0.05 | X |
| Terphenyls | 26140 - 60 - 3 | (C)1 | (C)9 | |
| 1,1,1,2-Tetrachloro-2,2-difluoroethane | 76 - 11 - 9 | 500 | 4170 | |
| 1,1,2,2-Tetrachloro-1,2-difluoroethane | 76 - 12 - 0 | 500 | 4170 | |
| 1,1,2,2-Tetrachloroethane | 79 - 34 - 5 | 5 | 35 | X |
| Tetrachloroethylene; see Perchloroethylene | | | | |
| Tetrachloromethane; see Carbon tetrachloride | | | | |
| Tetrachloronaphthalene | 1335 - 88 - 2 | | 2 | X |
| Tetraethyl lead (as Pb) | 78 - 00 - 2 | | 0.075 | X |
| Tetrahydrofuran | 109 - 99 - 9 | 200 | 590 | |
| Tetramethyl lead (as Pb) | 75 - 74 - 1 | | 0.075 | X |
| Tetramethyl succinonitrile | 3333 - 52 - 6 | 0.5 | 3 | X |
| Tetranitromethane | 509 - 14 - 8 | 1 | 8 | |
| Tetryl (2,4,6-Trinitrophenylmethylnitramine) | 479 - 45 - 8 | | 1.5 | X |
| Thallium, soluble compounds (as Tl) | 7440 - 28 - 0 | | 0.1 | X |
| 4,4'-Thiobis (6-tert, Butyl-m-cresol) | 96 - 69 - 5 | | | |
| Total dust | | | 15 | |
| Respirable fraction | | | 5 | |
| Thiram | 137 - 26 - 8 | | 5 | |
| Tin, inorganic compounds (except oxides) (as Sn) | 7440 - 31 - 5 | | 2 | |
| Tin, organic compounds (as Sn) | 7440 - 31 - 5 | | 0.1 | |
| Titanium dioxide | 13463 - 67 - 7 | | | |
| Total dust | | | 15 | |
| Toluene | 108 - 88 - 3 | | (2) | |
| Toluene-2,4-diisocyanate (TDI) | 584 - 84 - 9 | (C)0.02 | (C)0.14 | |
| o-Toluidine | 95 - 53 - 4 | 5 | 22 | X |
| Toxaphene; see Chlorinated camphene | | | | |
| Tremolite; see Silicates | | | | |
| Tributyl phosphate | 126 - 73 - 8 | | 5 | |
| 1,1,1-Trichloroethane; see Methyl chloroform | | | | |
| 1,1,2-Trichloroethane | 79 - 00 - 5 | 10 | 45 | X |
| Trichloroethylene | 79 - 01 - 6 | | (2) | |

| Substance | CAS No. (c) | ppm (a)1 | mg/m3 (b)1 | Skin designation |
|---|---|---|---|---|
| Trichloromethane; see Chloroform | | | | |
| Trichloronaphthalene | 1321 - 65 - 9 | | 5 | X |
| 1,2,3-Trichloropropane | 96 - 18 - 4 | 50 | 300 | |
| 1,1,2-Trichloro-1,2,2- trifluoroethane | 76 - 13 - 1 | 1000 | 7600 | |
| Triethylamine | 121 - 44 - 8 | 25 | 100 | |
| Trifluorobromomethane | 75 - 63 - 8 | 1000 | 6100 | |
| 2,4,6-Trinitrophenol; see Picric acid | | | | |
| 2,4,6-Trinitrophenylmethyl nitramine; see Tetryl | | | | |
| 2,4,6-Trinitrotoluene (TNT) | 118 - 96 - 7 | | 1.5 | X |
| Triorthocresyl phosphate | 78 - 30 - 8 | | 0.1 | |
| Triphenyl phosphate | 115 - 86 - 6 | | 3 | |
| Turpentine | 8006 - 64 - 2 | 100 | 560 | |
| Uranium (as U) | 7440 - 61 - 1 | | | |
| Soluble compounds | | | 0.05 | |
| Insoluble compounds | | | 0.25 | |
| Vanadium | 1314 - 62 - 1 | | | |
| Respirable dust (as $V_2O_5$) | | | (C)0.5 | |
| Fume (as $V_2O_5$) | | | (C)0.1 | |
| Vegetable oil mist | | | | |
| Total dust | | | 15 | |
| Respirable fraction | | | 5 | |
| Vinyl benzene; see Styrene | | | | |
| Vinyl chloride; see 1910.1017 | 75 - 01 - 4 | | | |
| Vinyl cyanide; see Acrylonitrile | | | | |
| Vinyl toluene | 25013 - 15 - 4 | 100 | 480 | |
| Warfarin | 81 - 81 - 2 | | 0.1 | |
| Xylenes (o-, m-, p-isomers) | 1330 - 20 - 7 | 100 | 435 | |
| Xylidine | 1300 - 73 - 8 | 5 | 25 | X |
| Yttrium | 7440 - 65 - 5 | | 1 | |
| Zinc chloride fume | 7646 - 85 - 7 | | 1 | |
| Zinc oxide fume | 1314 - 13 - 2 | | 5 | |
| Zinc oxide | 1314 - 13 - 2 | | | |
| Total dust | | | 15 | |
| Respirable fraction | | | 5 | |
| Zinc stearate | 557 - 05 - 1 | | | |
| Total dust | | | 15 | |
| Respirable fraction | | | 5 | |
| Zirconium compounds (as Zr) | 7440 - 67 - 7 | | 5 | |

[1]The PELs are 8-hour TWAs unless otherwise noted; a (C) designation denotes a ceiling limit. They are to be determined from breathing-zone air samples.

(a) Parts of vapor or gas per million parts of contaminated air by volume at 25°C and 760 torr.

(b) Milligrams of substance per cubic meter of air. When entry is in this column only, the value is exact; when listed with a ppm entry, it is approximate.

(c) The CAS number is for information only. Enforcement is based on the substance name. For an entry covering more than one metal compound, measured as the metal, the CAS number for the metal is given—not CAS numbers for the individual compounds.

(d) The final benzene standard in 1910.1028 applies to all occupational exposures to benzene except in some circumstances the distribution and sale of fuels, sealed containers and pipelines, coke production, oil and gas drilling and production, natural gas processing, and the percentage exclusion for liquid mixtures; for the excepted subsegments, the benzene limits in Table Z-2 apply. See 1910.1028 for specific circumstances.

(e)This 8-hour TWA applies to respirable dust as measured by a vertical elutriator cotton dust sampler or equivalent instrument. The time-weighted average applies to the cottom waste processing operations of waste recycling (sorting, blending, cleaning and willowing) and garnetting. See also 1910.1043 for cotton dust limits applicable to other sectors.

(f)All inert or nuisance dusts, whether mineral, inorganic, or organic, not listed specifically by substance name are covered by the Particulates Not Otherwise Regulated (PNOR) limit which is the same as the inert or nuisance dust limit of Table Z-3.

[2] See Table Z-2.

[3] See Table Z-3.

[4] Varies with compound.

## Table 4-4-2: Z-2.

| Substance | 8-hour time-weighted average | Acceptable ceiling concentration | Acceptable maximum peak above the acceptable ceiling concentration for an 8-hour shift | |
|---|---|---|---|---|
| | | | Concentration | Maximum duration |
| Benzene[a] (Z37.40 - 1969) | 10 ppm | 25 ppm | 50 ppm | 10 minutes |
| Beryllium and beryllium compounds (Z37.29 - 1970) | 2 mg/m³ | 5 mg/m³ | 25 mg/m³ | 30 minutes |
| Cadmium fume[b] (Z37.5 - 1970) | 0.1 mg/m³ | 0.3 mg/m³ | | |
| Cadmium dust[b] (Z37.5 - 1970) | 0.2 mg/m³ | 0.6 mg/m³ | | |
| Carbon disulfide (Z37.3 - 1968) | 20 ppm | 30 ppm | 100 ppm | 30 minutes |
| Carbon tetrachloride | 10 ppm | 25 ppm | 200 ppm | 5 min. in any 4 hrs |
| Chromic acid and chromates (Z37.7 - 1971) | | 1 mg/10m³ | | |
| Ethylene dibromide (Z37.31 - 1970) | 20 ppm | 30 ppm | 50 ppm | 5 minutes |
| Ethylene dichloride (Z37.21 - 1969) | 50 ppm | 100 ppm | 200 ppm | 5 min. in any 3 hrs |
| Fluoride as dust (Z37.28 - 1969) | 2.5 mg/m³ | | | |
| Formaldehyde; see 1910.1048 | | | | |
| Hydrogen fluoride (Z37.28 - 1969) | 3 ppm | | | |
| Hydrogen sulfide (Z37.2 - 1966) | | 20 ppm | 50 ppm | 10 mins. once, only if no other meas. exp. occurs |
| Mercury (Z37.8 - 1971) | | 1 mg/10m³ | | |
| Methyl chloride (Z37.18 - 1969) | 100 ppm | 200 ppm | 300 ppm | 5 mins. in any 3 hrs |
| Methylene Chloride: See §1919.52 | | | | |
| Organo (alkyl) mercury (Z37.30 - 1969) | 0.01 mg/m³ | 0.04 mg/m³ | | |
| Styrene (Z37.15 - 1969) | 100 ppm | 200 ppm | 600 ppm | 5 mins. in any 3 hrs |
| Tetrachloroethylene (Z37.22 - 1967) | 100 ppm | 200 ppm | 300 ppm | 5 mins. in any 3 hrs |
| Toluene (Z37.12 - 1967) | 200 ppm | 300 ppm | 500 ppm | 10 minutes |
| Trichloroethylene (Z37.19 - 1967) | 100 ppm | 200 ppm | 300 ppm | 5 mins. in any 2 hrs |

[a] This standard applies to the industry segments exempt from the 1 ppm 8-hour TWA and 5 ppm STEL of the benzene standard at 1910.1028.

[b] This standard applies to any operations or sectors for which the Cadmium standard, 1910.1027, is stayed or otherwise not in effect.

## Table 4-4-3: Z-3 Mineral Dusts

| Substance | mppcf (a) | mg/m³ |
|---|---|---|
| Silica: | | |
| Crystalline | | |
| Quartz (Respirable) | $\frac{250\ (b)}{\%\ SiO_2 + 5}$ | $\frac{10\ mg/m^3\ (e)}{\%\ SiO_2 + 2}$ |
| Quartz (Total Dust) | | $\frac{30\ mg/m^3}{\%\ SiO_2 + 2}$ |
| Cristobalite: Use 1/2 the value calculated from the count or mass formulae for quartz | | |
| Tridymite: Use 1/2 the value calculated from the formulae for quartz | | |
| Amorphous, including natural diatomaceous earth | 20 | $\frac{80\ mg/m^3}{\%\ SiO_2}$ |
| Silicates (less than 1% crystalline silica): | | |
| Mica | 20 | |
| Soapstone | 20 | |
| Talc (not containing asbestos) | 20 (c) | |
| Talc (containing asbestos) Use asbestos limit | | |
| Tremolite, asbestiform (see 29 CFR 1910.1001) | | |
| Portland cement | 50 | |
| Graphite (natural) | 15 | |
| Coal Dust: | | |
| Respirable fraction less than 5% $SiO_2$ | | 2.4 mg/m³ (e) |
| Respirable fraction greater than 5% $SiO_2$ | | $\frac{10\ mg/m^3\ (e)}{\%\ SiO_2 + 2}$ |
| Inert or Nuisance Dust: (d) | | |
| Respirable fraction | 15 | 5 mg/m³ |
| Total dust | 50 | 15 mg/m³ |

Note: Conversion factors - mppcf X 35.3 = million particles per cubic meter = particles per c.c.

[a] Millions of particles per cubic foot of air, based on impinger samples counted by light-field techniques.

[b] The percentage of crystalline silica in the formula is the amount determined from airborne samples, except in those instances in which other methods have been shown to be applicable.

[c] Containing less than 1% quartz; if 1% quartz or more, use quartz limit.

[d] All inert or nuisance dusts, whether mineral, inorganic, or organic, not listed specifically by substance name are covered by this limit, which is the same as the Particulates Not Otherwise Regulated (PNOR) limit in Table Z-1.

[e] Both concentration and percent quartz for the application of this limit are to be determined from the fraction passing a size-selector with the following characteristics:

| Aerodynamic diameter (unit density sphere) | Percent passing sector |
|---|---|
| 2 | 90 |
| 2.5 | 75 |
| 3.5 | 50 |
| 5.0 | 25 |
| 10 | 0 |

The measurements under this note refer to the use of an AEC (now NRC) instrument. The respirable fraction of coal dust is determined with an MRE; the figure corresponding to that of 2.4 mg/m³ in the table for coal dust is 4.5 mg/m³K.

[58 FR 35340, June 30. 1993; 58 FR 40191, July 27, 1993, as amended at 61 FR 56831, Nov. 4, 1996; 62 FR 1600, Jan. 10, 1997; 62 FR 42018, Aug. 4, 1997]

# Asbestos

## 1910.1001

Section 5 provides the basic guidelines for working with asbestos. It does not provide all the information the reader will need to control employee exposure; however, it does provide an overview for the reader to make informed decisions regarding employee exposure to asbestos. An industrial hygienist or abatement professional should be consulted if asbestos is suspected in your facility.

## General

- Asbestos includes chrysotile, amosite, crocidolite, tremolite asbestos, anthophyllite asbestos, actinolite asbestos, and any of these minerals that have been chemically treated and/or altered.
- Asbestos-containing material (ACM) means any material containing more than 1 percent asbestos.
- The employer shall ensure that no employee is exposed to an airborne concentration of asbestos in excess of 0.1 fiber per cubic centimeter of air (1 f/cc) as an 8-hour time-weighted average (TWA), or by an equivalent method.
- The employer shall ensure that no employee is exposed to an airborne concentration of asbestos in excess of 1.0 f/cc as averaged over a sampling period of 30 minutes.
- Determinations of employee exposure shall be made from breathing zone air samples that are representative of the 8-hour TWA and 30-minute short-term exposures of each employee.
- Representative 8-hour TWA employee exposures shall be determined on the basis of one or more samples representing full-shift exposures for each shift for each employee in each job classification in each work area. Representative 30-minute short-term employee exposures shall be determined on the basis of one or more samples representing 30-minute exposures associated

with operations that are most likely to produce exposures above the excursion limit for each shift for each job classification in each work area.

- ❑ Each employer shall perform initial monitoring of employees who are, or may reasonably be expected to be, exposed to airborne concentrations at or above the TWA permissible exposure limit and/or excursion limit.
- ❑ Where the employer has monitored for the TWA permissible exposure limit and/or the excursion limit, and the monitoring satisfies all other requirements, the employer may rely on such earlier monitoring results.
- ❑ No initial monitoring is required where the employer has relied upon objective data that demonstrate asbestos is not capable of being released in airborne concentrations at or above the TWA permissible exposure limit and/or excursion limit under the expected conditions of processing, use, or handling.
- ❑ After the initial determinations, samples shall be of such frequency and pattern as to represent with reasonable accuracy the levels of exposure of the employees. In no case shall sampling be at intervals greater than six months for employees whose exposures may reasonably be foreseen to exceed the TWA permissible exposure limit and/or excursion limit.
- ❑ If either the initial or the periodic monitoring statistically indicates that employee exposures are below the TWA permissible exposure limit and/or excursion limit, the employer may discontinue the monitoring for those employees whose exposures are represented by such monitoring.

## Exposure Monitoring

- ❑ The employer shall institute the exposure monitoring whenever there has been a change in the production, process, control equipment, personnel, or work practices that may result in new or additional exposures above the TWA permissible exposure limit and/or excursion limit, or when the employer has any reason to suspect that a change may result in new or additional exposures above the PEL and/or excursion limit.
- ❑ All samples taken to satisfy the monitoring shall be personal samples.
- ❑ All samples taken to satisfy the monitoring requirements shall be evaluated using the OSHA Reference Method (ORM) or an equivalent counting method.
- ❑ If an equivalent method to the ORM is used, the employer shall ensure that the method meets the following criteria:
    - Replicate exposure data used to establish equivalency are collected in side-by-side field and laboratory comparisons
    - The comparison indicates that 90 percent of the samples collected in the range 0.5 to 2.0 times the permissible limit have an accuracy range of plus or minus 25 percent of the ORM results at a 95 percent confidence level as demonstrated by a statistically valid protocol; and
    - The equivalent method is documented and the results of the comparison testing are maintained.

- To satisfy the monitoring requirements, employers must use the results of monitoring analysis performed by laboratories that have instituted quality assurance programs. (AIHA accredited labs are highly recommended.)

- The employer shall, within 15 working days after the receipt of the results of any monitoring, notify the affected employees of these results in writing either individually or by posting results in an appropriate location that is accessible to affected employees.

- The written notification shall contain the corrective action being taken by the employer to reduce employee exposure to or below the TWA and/or excursion limit, wherever monitoring results indicated that the TWA and/or excursion limit had been exceeded.

- The employer shall establish regulated areas wherever airborne concentrations of asbestos and/or PACM are in excess of the TWA and/or excursion limit.

- Regulated areas shall be demarcated from the rest of the workplace in any manner that minimizes the number of persons who will be exposed to asbestos.

- Access to regulated areas shall be limited to authorized persons.

- Each person entering a regulated area shall be supplied with, and required to use, a respirator.

- The employer shall ensure that employees do not eat, drink, smoke, chew tobacco or gum, or apply cosmetics in the regulated areas.

- The employer shall institute engineering controls and work practices to reduce and maintain employee exposure to or below the TWA and/or excursion limit.

- Wherever the feasible engineering controls and work practices that can be instituted are not sufficient to reduce employee exposure to or below the TWA and/or excursion limit, the employer shall use them to reduce employee exposure to the lowest levels achievable by these controls, and shall supplement them by the use of respiratory protection.

- Wherever feasible engineering controls and work practices that can be instituted are not sufficient to reduce the employee exposure to or below the TWA and/or excursion limit, the employer shall use them to reduce employee exposure to or below 0.5 fiber per cubic centimeter of air (as an 8-hour TWA) or 2.5 fibers/cc for 30 minutes (short-term exposure), and shall supplement them by the use of any combination of respiratory protection, work practices, and feasible engineering controls that will reduce employee exposure to or below the TWA and to or below the excursion limit, for the following operations:
  - Coupling cutoff in primary asbestos cement pipe manufacturing;
  - Sanding in primary and secondary asbestos cement sheet manufacturing;
  - Grinding in primary and secondary friction product manufacturing;
  - Carding and spinning in dry textile processes; and
  - Grinding and sanding in primary plastics manufacturing.

## Local Exhaust Ventilation

- ❑ Local exhaust ventilation and dust collection systems shall be designed, constructed, installed, and maintained in accordance with good practices, such as those found in the American National Standard Fundamentals Governing the Design and Operation of Local Exhaust Systems, ANSI Z9.2-1979.

- ❑ All hand- and power-operated tools which would produce or release fibers of asbestos, such as, but not limited to, saws, scorers, abrasive wheels, and drills, shall be provided with local exhaust ventilation systems.

- ❑ Asbestos shall be handled, mixed, applied, removed, cut, scored, or otherwise worked in a wet state sufficient to prevent the emission of airborne fibers so as to expose employees to levels in excess of the TWA and/or excursion limit.

- ❑ No asbestos cement, mortar, coating, grout, plaster, or similar material containing asbestos, shall be removed from bags, cartons, or other containers in which they are shipped, without being either wetted, enclosed, or ventilated so as to prevent effectively the release of airborne fibers.

- ❑ Compressed air shall not be used to remove asbestos or materials containing asbestos unless the compressed air is used in conjunction with a ventilation system that effectively captures the dust cloud created by the compressed air.

- ❑ Sanding of asbestos-containing flooring material is prohibited.

## Written Program

- ❑ Where the TWA and/or excursion limit is exceeded, the employer shall establish and implement a written program to reduce employee exposure to or below the TWA and to or below the excursion limit by means of engineering and work-practice controls, and by the use of respiratory protection.

- ❑ Such programs shall be reviewed and updated as necessary to reflect significant changes in the status of the employer's compliance program.

- ❑ Written programs shall be submitted upon request for examination and copying to the Assistant Secretary, the Director, affected employees, and designated employee representatives.

- ❑ The employer shall not use employee rotation as a means of compliance with the TWA and/or excursion limit.

- ❑ Specific compliance methods for brake and clutch repair:

  - During automotive brake and clutch inspection, disassembly, repair, and assembly operations, the employer shall institute engineering controls and work practices to reduce employee exposure to materials containing asbestos using a negative pressure enclosure/HEPA vacuum system method or low pressure/wet cleaning method.

- The employer may also comply by using an equivalent method which follows written procedures, which the employer demonstrates can achieve equivalent exposure reductions as do the two "preferred methods." Such demonstration must include monitoring data conducted under workplace conditions closely resembling the process, type of asbestos containing materials, control method, work practices and environmental conditions which the equivalent method will be used, or objective data which document that under all reasonably foreseeable conditions of brake and clutch repair applications.

## Respiratory Protection

- Respirators must be used during:
  - Periods necessary to install or implement feasible engineering and work-practice controls.
  - Work operations, such as maintenance and repair activities, for which engineering and work-practice controls are not feasible.
  - Work operations for which feasible engineering and work-practice controls are not yet sufficient to reduce employee exposure to or below the TWA and/or excursion limit.
  - Emergencies.
- The employer must implement a respiratory protection program in accordance with 29 CFR 1910.134 (b) through (d) (except (d)(1)(iii)), and (f) through (m).
- The employer must provide a tight-fitting, powered, air-purifying respirator instead of any negative-pressure respirator specified in Table 4-5-1 when an employee chooses to use this type of respirator and the respirator provides adequate protection to the employee.
- No employee must be assigned to tasks requiring the use of respirators if, based on their most recent medical examination, the examining physician determines that the employee will be unable to function normally using a respirator, or that the safety or health of the employee or other employees will be impaired by the use of a respirator.
- Such employees must be assigned to another job or given the opportunity to transfer to a different position, the duties of which they can perform. If such a transfer position is available, the position must be with the same employer, in the same geographical area, and with the same seniority, status, and rate of pay the employee had just prior to such transfer.
- The employer must select and provide the appropriate respirator from Table 4-5-1.

## Table 4-5-1: Respiratory Protection for Asbestos Fibers

| Airborne concentration of asbestos or conditions of use | Required respirator |
|---|---|
| Not in excess of 1 f/cc (10 X PEL) | Half-mask air-purifying respirator other than a disposable respirator, equipped with high-efficiency filters. |
| Not in excess of 5 f/cc (50 X PEL) | Full facepiece air-purifying respirator equipped with high-efficiency filters. |
| Not in excess of 10 f/cc (100 X PEL) | Any powered air-purifying respirator equipped with high-efficiency filters, or any supplied-air respirator operated in efficiency filters, or any supplied air respirator operated in continuous flow mode. |
| Not in excess of 100 f/cc (1,000 X PEL) | Full facepiece, supplied-air respirator operated in pressure-demand mode. |
| Greater than 100 f/cc (1,000 X PEL) or unknown concentration | Full facepiece, supplied-air respirator operated in pressure or demand mode, equipped with an auxiliary positive-pressure self-contained breathing apparatus. |

Notes: a. Respirators assigned for high environmental concentrations may be used at lower concentrations, or when required respirator use is independent of concentration.
b. A high-efficiency filter means a filter that is at least 99.97 percent efficient against mono-dispersed particles of 0.3 micrometers in diameter or larger.

## Protective Clothing and Equipment

- ❑ If an employee is exposed to asbestos above the TWA and/or excursion limit, or where the possibility of eye irritation exists, the employer shall provide at no cost to the employee, and ensure that the employee uses, appropriate protective work clothing and equipment, such as, but not limited to:
  - Coveralls or similar full-body work clothing;
  - Gloves, head coverings, and foot coverings; and
  - Face shields, vented goggles, or other appropriate protective equipment that complies with 29 CFR 1910.133.
- ❑ The employer shall ensure that employees remove work clothing contaminated with asbestos only in change rooms.
- ❑ The employer shall ensure that no employee takes contaminated work clothing out of the change room, except those employees authorized to do so for the purpose of laundering, maintenance, or disposal.
- ❑ Contaminated work clothing shall be placed and stored in closed containers that prevent dispersion of the asbestos outside the container.

- Containers of contaminated protective devices or work clothing which are to be taken out of change rooms or the workplace for cleaning, maintenance, or disposal, shall bear labels.
- The employer shall clean, launder, repair, or replace protective clothing and equipment to maintain their effectiveness. The employer shall provide clean protective clothing and equipment at least weekly to each affected employee.
- The employer shall prohibit the removal of asbestos from protective clothing and equipment by blowing or shaking.
- Laundering of contaminated clothing shall be done so as to prevent the release of airborne fibers of asbestos in excess of the permissible exposure limits.
- Any employer who gives contaminated clothing to another person for laundering shall inform such person to prevent effectively the release of airborne fibers of asbestos in excess of the permissible exposure limits.
- The employer shall inform any person who launders or cleans protective clothing or equipment contaminated with asbestos of the potentially harmful effects of exposure to asbestos.
- Contaminated clothing shall be transported in sealed impermeable bags, or other closed, impermeable containers, and labeled.
- The employer shall provide clean change rooms for employees who work in areas where their airborne exposure to asbestos is above the TWA and/or excursion limit.
- The employer shall ensure that change rooms are in accordance with 29 CFR 1910.141(e), and are equipped with two separate lockers or storage facilities, so separated as to prevent contamination of the employee's street clothes from his protective work clothing and equipment.
- The employer shall ensure that employees, who work in areas where their airborne exposure is above the TWA and/or excursion limit, shower at the end of the work shift.
- The employer shall provide shower facilities that comply with 29 CFR 1910.141(d)(3).
- The employer shall ensure that employees who are required to shower do not leave the work place wearing any clothing or equipment worn during the work shift.
- The employer shall provide lunchroom facilities for employees who work in areas where their airborne exposure is above the TWA and/or excursion limit.
- The employer shall ensure that lunchroom facilities have a positive-pressure, filtered air supply, and are readily accessible to employees.
- The employer shall ensure that employees who work in areas where their airborne exposure is above the PEL and/or excursion limit, wash their hands and faces prior to eating, drinking, or smoking.
- The employer shall ensure that employees do not enter lunchroom facilities with protective

work clothing or equipment unless surface asbestos fibers have been removed from the clothing or equipment by vacuuming or other method that removes dust without causing the asbestos to become airborne.

- The employer shall ensure that employees do not smoke in work areas where they are occupationally exposed to asbestos because of activities in that work area.

## General Industry

- Employers and building owners are required to treat installed TSI and sprayed on and troweled-on surfacing materials as ACM in buildings constructed before 1980. These materials are designated "presumed ACM or PACM."

- Asphalt and vinyl flooring material installed before 1980 also must be treated as asbestos-containing. The employer or building owner may demonstrate that PACM and flooring material do not contain asbestos.

- Building and facility owners shall determine the presence, location, and quantity of ACM and/or PACM at the worksite. Employers and building and facility owners shall exercise due diligence to inform employers and employees about the presence and location of ACM and PACM.

- Building and facility owners shall maintain records of all information known to the building owner concerning the presence, location, and quantity of ACM and PACM in the building/facility. Such records shall be kept for the duration of ownership and shall be transferred to successive owners.

- Building and facility owners shall inform employers, and employers shall inform employees who will perform housekeeping activities in areas which contain ACM and/or PACM, of the presence and location of ACM and/or PACM in such areas which may be contacted during such activities.

- At any time, an employer and/or building owner may demonstrate, for purposes of this standard, that PACM does not contain asbestos. Building owners and/or employers are not required to communicate information about the presence of building material for which such a demonstration has been made. However, in all such cases, the information, data and analysis supporting the determination that PACM does not contain asbestos, shall be retained.

- An employer or owner may demonstrate that PACM does not contain asbestos by:
  - Having a completed inspection conducted pursuant to the requirements of AHERA (40 CFR 763, Subpart E) which demonstrates that no ACM is present in the material; or
  - Performing tests of the material containing PACM which demonstrate that no ACM is present in the material. Such tests shall include analysis of bulk samples collected in the manner described in 40 CFR 763.86. The tests, evaluation, and sample collection shall be conducted by an accredited inspector or by a CIH. Analysis of samples shall be performed by persons or laboratories with proficiency demonstrated by current successful participation

in a nationally recognized testing program such as the National Voluntary Laboratory Accreditation Program (NVLAP), or the National Institute for Standards and Technology (NIST), or the Round Robin for bulk samples administered by the American Industrial Hygiene Association (AIHA), or an equivalent nationally recognized round robin testing program.

- The employer and/or building owner may demonstrate that flooring material, including associated mastic and backing, does not contain asbestos by the determination of an industrial hygienist based upon recognized analytical techniques showing that the material is not ACM.

## Communication of Hazards to Employees

- Warning signs shall be provided and displayed at each regulated area. In addition, warning signs shall be posted at all approaches to regulated areas so that an employee may read the signs and take necessary protective steps before entering the area.

- The warning signs shall look similar to the sample at above right.
- In addition, where the use of respirators and protective clothing is required in the regulated area, the warning signs shall look like the sample at below right.

- The employer shall ensure that employees working in and contiguous to regulated areas comprehend the warning signs. Means to ensure employee comprehension may include the use of foreign languages, pictographs, and graphics.
- At the entrance to mechanical rooms/areas in which employees reasonably can be expected to enter and which contain ACM and/or PACM, the building owner shall post signs which identify the material which is present, its location, and appropriate work practices which, if followed, will ensure that ACM and/or PACM will not be disturbed.
- The employer shall ensure, to the extent feasible, that employees who come in contact with these signs can comprehend them. Means to ensure employee comprehension may include the use of foreign languages, pictographs, graphics, and awareness training.
- Warning labels shall be affixed to all raw materials, mixtures, scrap, waste, debris, and other products containing asbestos fibers, or their containers. When a building owner or employer identifies previously installed ACM and/or PACM, labels or signs shall be affixed or posted so that employees will be notified of what materials contain ACM and/or PACM. The employer shall place such labels in areas where they will clearly be noticed by employees who are likely to be exposed, such as at the entrance to mechanical room/areas. Signs may be posted in lieu of labels so long as they contain information required for labeling.

- The labels shall comply with the requirements of 29 CFR 1910.1200(f) of OSHA's Hazard Communication Standard, and shall look similar to the sample at right.

DANGER

CONTAINS ASBESTOS FIBERS

AVOID CREATING DUST

CANCER AND LUNG DISEASE HAZARD

- Employers who are manufacturers or importers of asbestos or asbestos products shall comply with the requirements regarding development of material safety data sheets as specified in 29 CFR 1910.1200(g) of OSHA's Hazard Communication Standard.
- The provisions for labels or material safety data sheets do not apply where:
  - Asbestos fibers have been modified by a bonding agent, coating, binder, or other material provided that the manufacturer can demonstrate that during any reasonably foreseeable use, handling, storage, disposal, processing, or transportation, no airborne concentrations of asbestos fibers in excess of the TWA permissible exposure level and/or excursion limit will be released; or
  - Asbestos is present in a product in concentrations of less than 1.0 percent.
- The employer shall institute a training program for all employees who are exposed to airborne concentrations of asbestos at or above the PEL and/or excursion limit and ensure their participation in the program.
- Training shall be provided prior to or at the time of initial assignment and at least annually thereafter.
- The training program shall be conducted in a manner which the employee is able to understand. The employer shall ensure that each employee is informed of the following:
  - The health effects associated with asbestos exposure;
  - The relationship between smoking, exposure to asbestos, and lung cancer;
  - The quantity, location, manner of use, release, and storage of asbestos, and the specific nature of operations which could result in exposure to asbestos;
  - The engineering controls and work practices associated with the employee's job assignment;
  - The specific procedures implemented to protect employees from exposure to asbestos, such as appropriate work practices, emergency and clean up procedures, and personal protective equipment to be used;
  - The purpose, proper use, and limitations of respirators and protective clothing, if appropriate;
  - The purpose and a description of the medical surveillance program;
  - The content of OSHA's Hazard Communication Standard (29 CFR 1910.1200), including appendices;

- The names, addresses, and phone numbers of public health organizations which provide information, materials, and/or conduct programs concerning smoking cessation. The employer may distribute the list of such organizations to comply with this requirement; and
- The requirements for posting signs and affixing labels and the meaning of the required legends for such signs and labels.

❑ The employer shall also provide, at no cost to employees who perform housekeeping operations in an area which contains ACM or PACM, an asbestos awareness training course, which shall at a minimum contain the following elements:

- Health effects of asbestos;
- Locations of ACM and PACM in the building or facility;
- Recognition of ACM and PACM damage and deterioration;
- Requirements in this standard relating to housekeeping; and
- Proper response to fiber release episodes.

❑ Each such employee shall receive refresher training at least once a year.

❑ The employer shall make a copy of 29 CFR 1910.1001 and its appendices readily available without cost to all affected employees.

❑ The employer shall provide, upon request, all materials relating to the employee information and training program to the Assistant Secretary and the Director.

❑ The employer shall inform all employees of the availability of self-help smoking cessation program material. Upon employee request, the employer shall distribute such material, consisting of NIH Publication No. 89-1647, or equivalent self-help material, which is approved or published by a public health organization.

## Housekeeping

❑ All surfaces shall be maintained as free as practicable of ACM waste and debris and accompanying dust.

❑ All spills and sudden releases of material containing asbestos shall be cleaned up as soon as possible.

❑ Surfaces contaminated with asbestos may not be cleaned by the use of compressed air.

❑ HEPA-filtered vacuuming equipment shall be used for vacuuming asbestos containing waste and debris. The equipment shall be used and emptied in a manner which minimizes the reentry of asbestos into the workplace.

❑ Waste, scrap, debris, bags, containers, equipment, and clothing contaminated with asbestos consigned for disposal; shall be collected, recycled, and disposed of in sealed, impermeable bags, or other closed, impermeable containers.

- ❑ Shoveling, dry sweeping, and dry cleanup of asbestos may be used only where vacuuming and/or wet cleaning are not feasible.

- ❑ Sanding of asbestos-containing floor material is prohibited.

- ❑ Stripping of finishes shall be conducted using wet methods and low abrasion pads at speeds lower than 300 rpm.

- ❑ Burnishing or dry buffing may only be performed on asbestos-containing flooring which has sufficient finish so that the pad cannot contact the asbestos-containing material.

- ❑ Waste and debris and accompanying dust in an area containing accessible ACM and/or PACM or visibly deteriorated ACM, shall not be dusted, swept dry, or vacuumed without using a HEPA filter.

## Medical Surveillance

- ❑ The employer shall institute a medical surveillance program for all employees who are or will be exposed to airborne concentrations of fibers of asbestos at or above the TWA and/or excursion limit.

- ❑ The employer shall ensure that all medical examinations and procedures are performed by or under the supervision of a licensed physician, and shall be provided without cost to the employee and at a reasonable time and place.

- ❑ Persons who administer the pulmonary function testing, other than licensed physicians, shall complete a training course in spirometry sponsored by an appropriate academic or professional institution.

- ❑ Before an employee is assigned to an occupation exposed to airborne concentrations of asbestos fibers at or above the TWA and/or excursion limit, a pre-placement medical examination shall be provided or made available by the employer.

- ❑ Such examination shall include, at a minimum, a medical and work history; a complete physical examination of all systems with emphasis on the respiratory system, the cardiovascular system, and digestive tract; completion of the respiratory disease standardized questionnaire; a chest roentgenogram (posterior-anterior 14 x 17 inches); pulmonary function tests to include forced vital capacity (FVC) and forced expiratory volume at 1 second (FEV(1.0)); and any additional tests deemed appropriate by the examining physician. Interpretation and classification of chest roentgenogram shall be conducted.

- ❑ Periodic medical examinations shall be made available annually.

- ❑ The scope of the medical examination shall be in conformance with the protocol except that the frequency of chest roentgenogram shall be conducted in accordance with Table 4-5-2, and the abbreviated standardized questionnaire shall be administered to the employee.

## Table 4-5-2: Frequency of Chest Roentgenogram

| Years since first exposure | Age of employee | | |
|---|---|---|---|
| | 15 to 35 | 35 to 45 | 45+ |
| 0 to 10 | Every 5 years | Every 5 years | Every 5 years |
| 10+ | Every 5 years | Every 2 years | Every 1 year |

- The employer shall provide, or make available, a termination of employment medical examination for any employee who has been exposed to airborne concentrations of fibers of asbestos at or above the TWA and/or excursion limit.

- The medical examination shall be in accordance with the requirements of the periodic examinations and shall be given within 30 calendar days before or after the date of termination of employment.

- No medical examination is required of any employee, if adequate records show that the employee has been examined within the past year. A pre-employment medical examination that was required as a condition of employment by the employer may not be used by that employer to meet the requirements of this paragraph, unless the cost of such examination is borne by the employer.

- The employer shall provide the following information to the examining physician:
  - A copy of 29 CFR 1910.1001 and Appendices D and E;
  - A description of the affected employee's duties as they relate to the employee's exposure;
  - The employee's representative exposure level or anticipated exposure level;
  - A description of any personal protective and respiratory equipment used or to be used; and
  - Information from previous medical examinations of the affected employee that is not otherwise available to the examining physician.

- The employer shall obtain a written, signed opinion from the examining physician. This written opinion shall contain the results of the medical examination and shall include:
  - The physician's opinion as to whether the employee has any detected medical conditions that would place the employee at an increased risk of material health impairment from exposure to asbestos;
  - Any recommended limitations on the employee or upon the use of personal protective equipment such as clothing or respirators;
  - A statement that the employee has been informed by the physician of the results of the medical examination and of any medical conditions resulting from asbestos exposure that require further explanation or treatment; and
  - A statement that the employee has been informed by the physician of the increased risk of lung cancer attributable to the combined effect of smoking and asbestos exposure.

- ❑ The employer shall instruct the physician not to reveal in the written opinion given to the employer specific findings or diagnoses unrelated to occupational exposure to asbestos.

- ❑ The employer shall provide a copy of the physician's written opinion to the affected employee within 30 days of its receipt.

## Recordkeeping

- ❑ The employer may utilize the services of competent organizations, such as industry trade associations and employee associations, to maintain the records.

- ❑ The employer shall keep an accurate record of all measurements taken to monitor employee exposure to asbestos.

- ❑ This record shall include at least the following information:
  - The date of measurement;
  - The operation involving exposure to asbestos which is being monitored;
  - Sampling and analytical methods used and evidence of their accuracy;
  - Number, duration, and results of samples taken;
  - Type of respiratory protective devices worn, if any; and
  - Name, social security number, and exposure of the employees represented.

- ❑ The employer shall maintain this record for at least 30 years, in accordance with 29 CFR 1910.20.

- ❑ Where the processing, use, or handling of products made from or containing asbestos is exempted, the employer shall establish and maintain an accurate record of objective data reasonably relied upon in support of the exemption.

- ❑ The record shall include at least the following:
  - The product qualifying for exemption;
  - The source of the objective data;
  - The testing protocol, results of testing, and/or analysis of the material for the release of asbestos;
  - A description of the operation exempted and how the data support the exemption; and
  - Other data relevant to the operations, materials, processing, or employee exposures covered by the exemption.

- ❑ The employer shall maintain this record for the duration of the employer's reliance upon such objective data.

- ❑ The employer shall establish and maintain an accurate record for each employee subject to medical surveillance in accordance with 29 CFR 1910.20.

- The record shall include at least the following information:
  - The name and social security number of the employee;
  - The physician's written opinions;
  - Any employee medical complaints related to exposure to asbestos; and
  - A copy of the information provided to the physician.
- The employer shall ensure that this record is maintained for the duration of employment plus 30 years, in accordance with 29 CFR 1910.20.
- The employer shall maintain all employee training records for one year beyond the last date of employment of that employee.
- The employer, upon request, shall make all records available to the Assistant Secretary and the Director for examination and copying.
- The employer, upon request, shall make any exposure records available for examination and copying to affected employees, former employees, designated representatives, and the Assistant Secretary, in accordance with 29 CFR 1910.20 (a) through (e), and (g) through (i).
- The employer, upon request, shall make employee medical records available for examination and copying to the subject employee, anyone having the specific written consent of the subject employee, and the Assistant Secretary, in accordance with 29 CFR 1910.20.
- The employer shall comply with the requirements concerning transfer of records set forth in 29 CFR 1910.20(h).
- Whenever the employer ceases to do business and there is no successor employer to receive and retain the records for the prescribed period, the employer shall notify the Director at least 90 days prior to disposal of records and, upon request, transmit them to the Director.

## Observation of Monitoring

- The employer shall provide affected employees or their designated representatives an opportunity to observe any monitoring of employee exposure to asbestos.
- When observation of the monitoring of employee exposure to asbestos requires entry into an area where the use of protective clothing or equipment is required, the observer shall be provided with, and be required to use, such clothing and equipment and shall comply with all other applicable safety and health procedures.

# Benzene

## 1910.1028

Section 6 applies to all occupational exposures to benzene. This section does not apply to the storage, transportation, distribution, dispensing, sale, or use of gasoline, motor fuels, or other fuels containing benzene subsequent to its final discharge from bulk wholesale storage facilities, except that operations where gasoline or motor fuels are dispensed for more than four hours per day in an indoor location.

## Definitions

- *Action level* means an airborne concentration of benzene of 0.5 parts per million (ppm) calculated as an 8-hour time-weighted average (TWA).
- *Assistant Secretary* means the Assistant Secretary of Labor for Occupational Safety and Health, U.S. Department of Labor, or designee.
- *Authorized person* means any person specifically authorized by the employer whose duties require the person to enter a regulated area, or any person entering such an area as a designated representative of employees for the purpose of exercising the right to observe monitoring and measuring procedures.
- *Benzene* ($C_6H_6$) (CAS No. 71-43-2) means liquefied or gaseous benzene. It includes benzene contained in liquid mixtures and the benzene vapors released by these liquids. It does not include trace amounts of unreacted benzene contained in solid materials.
- *Bulk wholesale storage facility* means a bulk terminal or bulk plant where fuel is stored prior to its delivery to wholesale customers.
- *Container* means any barrel, bottle, can, cylinder, drum, reaction vessel, storage tank, or the like, but does not include piping systems.

- *Day* means any part of a calendar day.

- *Director* means the Director of the National Institute for Occupational Safety and Health, U.S. Department of Health and Human Services, or designee.

- *Emergency* means any occurrence such as, but not limited to, equipment failure, rupture of containers, or failure of control equipment which may or does result in an unexpected significant release of benzene.

- *Employee exposure* means exposure to airborne benzene which would occur if the employee were not using respiratory protective equipment.

- *Regulated area* means any area where airborne concentrations of benzene exceed or can reasonably be expected to exceed, the permissible exposure limits, either the 8-hour TWA exposure of 1 ppm or the short-term exposure limit of 5 ppm for 15 minutes.

- Vapor control system means any equipment used for containing the total vapors displaced during the loading of gasoline, motor fuel, or other fuel tank trucks and the displacing of these vapors through a vapor processing system or balancing the vapor with the storage tank. This equipment also includes systems containing the vapors displaced from the storage tank during the unloading of the tank truck which balance the vapors back to the tank truck.

## Permissible Exposure Limits (PELs)

- The employer shall assure that no employee is exposed to an airborne concentration of benzene in excess of 1 part of benzene per million parts of air (1 ppm) as an 8-hour TWA.

- The employer shall assure that no employee is exposed to an airborne concentration of benzene in excess of 5 ppm as averaged over any 15-minute period.

## Regulated Areas

- The employer shall establish a regulated area wherever the airborne concentration of benzene exceeds or can reasonably be expected to exceed the PELs, either the 8-hour TWA exposure of 1 ppm or the STEL limit of 5 ppm for 15 minutes.

- Access to regulated areas shall be limited to authorized persons.

- Regulated areas shall be determined from the rest of the workplace in any manner that minimizes the number of employees exposed to benzene within the regulated area.

## Exposure Monitoring

- Determinations of employee exposure shall be made from breathing zone air samples that are representative of each employee's average exposure to airborne benzene.

- Representative 8-hour TWA employee exposures shall be determined on the basis of one sample or samples representing the full shift exposure for each job classification in each work area.

- Determinations of compliance with the STEL shall be made from 15-minute employee breathing zone samples measured at operations where there is reason to believe exposures are high, such as where tanks are opened, filled, unloaded, or gauged; where containers or process equipment are opened; and where benzene is used for cleaning or as a solvent in an uncontrolled situation. The employer may use objective data, such as measurements from brief period measuring devices, to determine where STEL monitoring is needed.

- Except for initial monitoring, where the employer can document that one shift will consistently have higher employee exposures for an operation, the employer shall only be required to determine representative employee exposure for that operation during the shift on which the highest exposure is expected.

- Each employer who has a place of employment shall monitor each of these workplaces and work operations to determine accurately the airborne concentrations of benzene to which employees may be exposed.

- The initial monitoring shall be completed within 30 days of the introduction of benzene into the workplace.

- If the monitoring reveals employee exposure at or above the action level, but at or below the TWA, the employer shall repeat such monitoring for each such employee at least every year.

- If the monitoring reveals employee exposure above the TWA, the employer shall repeat such monitoring for each such employee at least every six months.

- The employer may alter the monitoring schedule from every six months to annually for any employee for whom two consecutive measurements taken at least seven days apart indicate that the employee exposure has decreased to the TWA or below, but is at or above the action level.

- Monitoring for the STEL shall be repeated as necessary to evaluate exposures of employees subject to short-term exposures.

- If the initial monitoring reveals employee exposure to be below the action level, the employer may discontinue the monitoring for that employee.

- If the periodic monitoring reveals that employee exposures, as indicated by at least two consecutive measurements taken at least seven days apart, are below the action level, the employer may discontinue the monitoring for that employee.

- The employer shall institute the exposure monitoring when there has been a change in the production, process, control equipment, personnel, or work practices which may result in new or additional exposures to benzene, or when the employer has any reason to suspect a change which may result in new or additional exposures.

- ❑ Whenever spills, leaks, ruptures, or other breakdowns occur that may lead to employee exposure, the employer shall monitor (using area or personal sampling) after the cleanup of the spill or repair of the leak, rupture, or other breakdown to ensure that exposures have returned to the level that existed prior to the incident.

- ❑ Monitoring shall be accurate, to a confidence level of 95 percent, to within plus or minus 25 percent for airborne concentrations of benzene.

- ❑ The employer shall, within 15 working days after the receipt of the results of any monitoring, notify each employee, in writing, either individually or by posting results in an appropriate location that is accessible to all affected employees.

- ❑ Whenever the PELs are exceeded, the written notification shall contain the corrective action being taken by the employer to reduce the employee exposure to or below the PEL, or shall refer to a document available to the employee which states the corrective actions to be taken.

## Methods of Compliance

- ❑ The employer shall institute engineering controls and work practices to reduce and maintain employee exposure to benzene at or below the permissible exposure limits, except to the extent that the employer can establish that these controls are not feasible.

- ❑ Wherever the feasible engineering controls and work practices which can be instituted are not sufficient to reduce employee exposure to or below the PELs, the employer shall use them to reduce employee exposure to the lowest levels achievable by these controls and shall supplement them by the use of respiratory protection.

- ❑ Where the employer can document that benzene is used in a workplace less than a total of 30 days per year, the employer shall use engineering controls, work-practice controls, respiratory protection, or any combination of these controls to reduce employee exposure to benzene to or below the PELs, except that employers shall use engineering and work-practice controls, if feasible, to reduce exposure to or below 10 ppm as an 8-hour TWA.

- ❑ When any exposures are over the PEL, the employer shall establish and implement a written program to reduce employee exposure to or below the PEL primarily by means of engineering and work-practice controls.

- ❑ The written program shall include a schedule for development and implementation of the engineering and work-practice controls. These plans shall be reviewed and revised as appropriate, based on the most recent exposure monitoring data, to reflect the current status of the program.

- ❑ Written compliance programs shall be furnished upon request for examination and copying to the Assistant Secretary, the Director, affected employees, and designated employee representatives.

## Respiratory Protection

- The employer must provide respirators to be used during:
  - Periods necessary to install or implement feasible engineering and work-practice controls.
  - Work operations for which the employer establishes that compliance with either the TWA or STEL through the use of engineering and work-practice controls is not feasible; for example, some maintenance and repair activities, vessel cleaning, or other operations for which engineering and work-practice controls are infeasible because exposures are intermittent and limited in duration.
  - Work operations for which feasible engineering and work-practice controls are not yet sufficient to reduce employee exposure to or below the PELs.
  - Emergencies.
- The employer must implement a respiratory protection program in accordance with 29 CFR 1910.134 (b) through (d) (except (d)(1)(iii) and (d)(3)(iii)(B)(1) and (2)); and (f) through (m).
- For air-purifying respirators, the employer must replace the air-purifying element at the expiration of its service life or at the beginning of each shift in which such elements are used, whichever comes first.
- If NIOSH approves an air-purifying element with an end-of-service-life indicator for benzene, such an element may be used until the indicator shows no further useful life.
- The employer must select the appropriate respirator from Table 4-6-1.
- Any employee who cannot use a negative-pressure respirator must be allowed to use a respirator with less breathing resistance, such as a powered air-purifying respirator or supplied-air respirator.

## Table 4-6-1: Respiratory Protection for Benzene

| Airborne concentration of benzene or condition of use | Respirator type |
|---|---|
| Less than or equal to 10 ppm | Half-mask air-purifying respirator with organic vapor cartridge. |
| Less than or equal to 50 ppm | Full facepiece respirator with organic vapor cartridge; or |
| | Full facepiece gas mask with chin style canister.[1] |
| Less than or equal to 100 ppm | Full facepiece powered air-purifying respirator with organic vapor canister.[1] |
| Less than or equal to 1,000 ppm | Supplied-air respirator with full facepiece in positive-pressure mode. |
| Greater than 1,000 ppm or unknown concentration | Self-contained breathing apparatus with full facepiece in positive-pressure mode; or |
| | Full facepiece positive-pressure supplied-air respirator with auxiliary self-contained air supply. |
| Escape | Any organic vapor gas mask; or |
| | Any self-contained breathing apparatus with full facepiece. |
| Firefighting | Full facepiece self-contained breathing apparatus in positive-pressure mode. |

[1] Canisters must have a minimum service life of four hours when tested at 150 ppm benzene, at a flow rate of 64 LPM, 25°C, and 85 percent relative humidity for non-powered air purifying respirators. The flow rate shall be 115 LPM and 170 LPM respectively for tight-fitting and loose-fitting powered air-purifying respirators.

## Protective Clothing and Equipment

- ❑ Personal protective clothing and equipment shall be worn where appropriate to prevent eye contact and to limit dermal exposure to liquid benzene. The employer shall provide protective clothing and equipment to the employee at no cost, and the employer shall assure its use where appropriate. Eye and face protection shall meet the requirements of 29 CFR 1910.133.

## Medical Surveillance

- ❑ The employer shall make available a medical surveillance program for employees who are or may be exposed to benzene at or above the action level 30 or more days per year; for employees who are or may be exposed to benzene at or above the PELs 10 or more days per year; and for employees involved in the tire building operations, called tire building machine operators, who use solvents containing greater than 0.1 percent benzene.

- ❑ The employer shall assure that all medical examinations and procedures are performed by or under the supervision of a licensed physician and that an accredited laboratory conducts all laboratory tests.

- The employer shall assure that persons, other than licensed physicians who administer the pulmonary function testing required by this section, shall complete a training course in spirometry sponsored by an appropriate governmental, academic, or professional institution.

- The employer shall assure that all examinations and procedures are provided without cost to the employee and at a reasonable time and place.

- The employer shall provide each employee with a medical examination which includes:
  - A detailed occupational history which includes:
    - Past work exposure to benzene or any other hematological toxins;
    - A family history of blood dyscrasias, including hematological neoplasms;
    - A history of blood dyscrasias, including genetic hemoglobin abnormalities, bleeding abnormalities, and abnormal function of formed blood elements;
    - A history of renal or liver dysfunction;
    - A history of medicinal drugs routinely taken;
    - A history of previous exposure to ionizing radiation; and
    - Exposure to marrow toxins outside of the current work situation.
  - A complete physical examination.
  - A complete blood count, including a leukocyte count with differential, a quantitative thrombocyte count, hematocrit, hemoglobin, erythrocyte count, and erythrocyte indices (MCV, MCH, MCHC). The examining physician shall review the results of these tests.
  - Additional tests as necessary in the opinion of the examining physician, based on alterations to the components of the blood or other signs which may be related to benzene exposure.
  - Special attention to the cardiopulmonary system and a pulmonary function test for all workers who are required to wear respirators for at least 30 days a year.

- No initial medical examination is required if adequate records show that the employee has been examined in accordance with the procedures.

- The employer shall provide each employee with a medical examination annually following the previous examination. These periodic examinations shall include at least the following elements:

  - A brief history regarding any new exposure to potential marrow toxins, changes in medicinal drug use, and the appearance of physical signs relating to blood disorders;
  - A complete blood count, including a leukocyte count with differential, quantitative thrombocyte count, hemoglobin, hematocrit, erythrocyte count, and erythrocyte indices (MCV, MCH, MCHC); and
  - Appropriate additional tests as necessary, in the opinion of the examining physician, in consequence of alterations in the components of the blood or other signs which may be related to benzene exposure.

- ❑ Where the employee develops signs and symptoms commonly associated with toxic exposure to benzene, the employer shall provide the employee with an additional medical examination that shall include those elements considered appropriate by the examining physician.

- ❑ For persons required to use respirators for at least 30 days a year, a pulmonary function test shall be performed every three years. A specific evaluation of the cardiopulmonary system shall be made at the time of the pulmonary function test.

- ❑ In addition to the surveillance, if an employee is exposed to benzene in an emergency situation, the employer shall have the employee provide a urine sample at the end of the employee's shift and have a urinary phenol test performed on the sample within 72 hours. The urine-specific gravity shall be corrected to 1.024.

- ❑ If the result of the urinary phenol test is below 75 mg phenol/L of urine, no further testing is required.

- ❑ If the result of the urinary phenol test is equal to or greater than 75 mg phenol/L of urine, the employer shall provide the employee with a complete blood count, including an erythrocyte count, leukocyte count with differential and thrombocyte count, at monthly intervals for a duration of three months following the emergency exposure.

- ❑ The employer shall provide the employees with periodic examinations if directed by the physician.

- ❑ The blood count shall be repeated within two weeks and the results of the complete blood count required for the initial and periodic examinations shall indicate if any of the following abnormal conditions exist:

  - The hemoglobin level or the hematocrit falls below the normal limit [outside the 95 precent confidence interval (C.I.)] as determined by the laboratory for the particular geographic area and/or these indices show a persistent downward trend from the individual's pre-exposure norms, provided these findings cannot be explained by other medical reasons.
  - The thrombocyte (platelet) count varies more than 20 percent below the employee's most recent values or falls outside the normal limit (95 percent C.I.) as determined by the laboratory.
  - The leukocyte count is below 4,000 per $mm^3$ or there is an abnormal differential count.

- ❑ If the abnormality persists, the examining physician shall refer the employee to a hematologist or an internist for further evaluation unless the physician has good reason to believe such referral is unnecessary.

- ❑ The employer shall provide the hematologist or internist with the information required to be provided to the physician and the medical record required to be maintained.

- ❑ The hematologist's or internist's evaluation shall include a determination as to the need for additional tests, and the employer shall assure that these tests are provided.

- ❑ The employer shall provide the following information to the examining physician:

- A copy of 29 CFR 1910.1028 and its appendices;
- A description of the affected employee's duties as they relate to the employee's exposure;
- The employee's actual or representative exposure level:
- A description of any personal protective equipment used, or to be used; and
- Information from previous employment-related medical examinations of the affected employee which is not otherwise available to the examining physician.

- ❑ For each examination under this section, the employer shall obtain and provide the employee with a copy of the examining physician's written opinion within 15 days of the examination. The written opinion shall be limited to the following information:
  - The occupationally pertinent results of the medical examination and tests;
  - The physician's opinion concerning whether the employee has any detected medical conditions which would place the employee's health at greater than normal risk of material impairment from exposure to benzene;
  - The physician's recommended limitations upon the employee's exposure to benzene or upon the employee's use of protective clothing, equipment, and respirators; and
  - A statement that the employee has been informed by the physician of the results of the medical examination and any medical conditions resulting from benzene exposure which require further explanation or treatment.

- ❑ The written opinion obtained by the employer shall not reveal specific records, findings and diagnoses that have no bearing on the employee's ability to work in a benzene-exposed work place.

- ❑ When a physician makes a referral to a hematologist/internist, the employee shall be removed from areas where exposures may exceed the action level until such time as the physician makes a determination.

- ❑ Following the examination and evaluation by the hematologist/internist, a decision to remove an employee from areas where benzene exposure is above the action level or to allow the employee to return to areas where benzene exposure is above the action level shall be made by the physician in consultation with the hematologist/internist. This decision shall be communicated in writing to the employer and employee. In the case of removal, the physician shall state the required probable duration of removal from occupational exposure to benzene above the action level and the requirements for future medical examinations to review the decision.

- ❑ For any employee who is removed, the employer shall provide a follow-up examination. The physician, in consultation with the hematologist/internist, shall make a decision within six months of the date the employee was removed as to whether the employee shall be returned to the usual job or whether the employee should be removed permanently.

- ❑ Whenever an employee is temporarily removed from benzene exposure, the employer shall transfer the employee to a comparable job for which the employee is qualified (or can be trained for in a short period) and where benzene exposures are as low as possible, but in no event higher than the action level. The employer shall maintain the employee's current wage rate, seniority, and other benefits. If there is no such job available, the employer shall provide medical removal protection benefits until such a job becomes available, or for six months, whichever comes first.

- ❑ Whenever an employee is removed permanently from benzene exposure based on a physician's recommendation, the employee shall be given the opportunity to transfer to another position which is available or later becomes available, for which the employee is qualified (or can be trained for in a short period) and where benzene exposures are as low as possible, but in no event higher than the action level. The employer shall assure that such employee suffers no reduction in current wage rate, seniority, or other benefits as a result of the transfer.

- ❑ The employer shall provide six months of medical removal protection benefits to an employee immediately following each occasion an employee is removed from exposure to benzene because of hematological findings, unless the employee has been transferred to a comparable job where benzene exposures are below the action level.

- ❑ For the purposes of this section, the requirement that an employer provide medical removal protection benefits means that the employer shall maintain the current wage rate, seniority, and other benefits of an employee as though the employee had not been removed.

- ❑ The employer's obligation to provide medical removal protection benefits to a removed employee shall be reduced to the extent that the employee receives compensation for earnings lost during the period of removal either from a publicly- or employer-funded compensation program, or from employment with another employer made possible by virtue of the employee's removal.

## Communication of Hazards to Employees

- ❑ The employer shall post signs at entrances to regulated areas. The signs shall look like the sample at right.

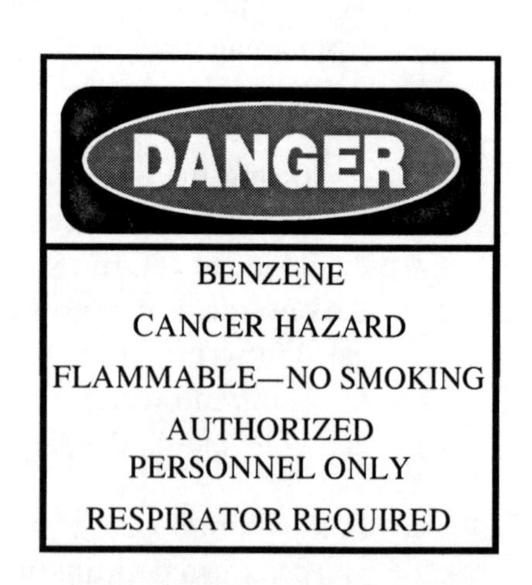


- ❑ The employer shall ensure that labels or other appropriate forms of warning are provided for containers of benzene within the workplace. There is no requirement to label pipes. The labels shall comply with the requirements of 29 CFR 1910.1200(f), and shall look like the sample at below right.

- ❑ Employers shall obtain or develop, and provide access for employees to a material safety data sheet (MSDS) which addresses benzene and complies with 29 CFR 1910.1200.

- ❑ Employers who are manufacturers or importers shall comply with the requirement in OSHA's Hazard Communication Standard, 29 CFR 1910.1200, to deliver to downstream employers an MSDS that addresses benzene.

- ❑ The employer shall provide employees with information and training at the time of their initial assignment to a work area where benzene is present. If exposures are above the action level, employees shall be provided with information and training at least annually thereafter.

- The training program shall be in accordance with the requirements of 29 CFR 1910.1200(h)(1) and (2), and shall include specific information on benzene for each category of information included in that section.

- In addition to the information required under 29 CFR 1910.1200, the employer shall describe the medical surveillance program.

## Recordkeeping

- The employer shall establish and maintain an accurate record of all measurements, in accordance with 29 CFR 1910.20.

- This record shall include:
  - The dates, number, duration, and results of each of the samples taken, including a description of the procedure used to determine representative employee exposures;
  - A description of the sampling and analytical methods used;
  - A description of the type of respiratory protective devices worn, if any; and
  - The name, social security number, job classification, and exposure levels of the employee monitored and all other employees whose exposure the measurement is intended to represent.

- The employer shall maintain this record for at least 30 years, in accordance with 29 CFR 1910.20.

- The employer shall establish and maintain an accurate record for each employee subject to medical surveillance, in accordance with 29 CFR 1910.20.

- This record shall include:
  - The name and social security number of the employee;
  - The employer's copy of the physician's written opinion on the initial, periodic, and special examinations, including results of medical examinations and all tests, opinions, and recommendations;
  - Any employee medical complaints related to exposure to benzene;
  - A copy of the information provided to the physician; and
  - A copy of the employee's medical and work history related to exposure to benzene or any other hematologic toxins.

- The employer shall maintain this record for at least the duration of employment plus 30 years, in accordance with 29 CFR 1910.20.

- ❑ The employer shall assure that all records required to be maintained by this section shall be made available, upon request, to the Assistant Secretary and the Director for examination and copying.

- ❑ Employee-exposure monitoring records required by this paragraph shall be provided upon request for examination and copying to employees, employee representatives, and the Assistant Secretary in accordance with 29 CFR 1910.20 (a) through (e) and (g) through (i).

- ❑ Employee medical records required by this paragraph shall be provided upon request for examination and copying to the subject employee, anyone having the specific written consent of the subject employee, and the Assistant Secretary, in accordance with 29 CFR 1910.20.

- ❑ The employer shall comply with the requirements involving transfer of records set forth in 29 CFR 1019.20(h).

- ❑ If the employer ceases to do business and there is no successor employer to receive and retain the records for the prescribed period, the employer shall notify the Director at least three months prior to disposal, and transmit the records to the Director if required by the Director within that period.

## Observation of Monitoring

- ❑ The employer shall provide affected employees, or their designated representatives, an opportunity to observe the measuring or monitoring of employee exposure to benzene.

- ❑ When observation of the measuring or monitoring of employee exposure to benzene requires entry into areas where the use of protective clothing and equipment or respirators is required, the employer shall provide the observer with personal protective clothing and equipment and respirators required to be worn by employees working in the area; assure the use of such clothing and equipment or respirators; and require the observer to comply with all other applicable safety and health procedures.

# Cadmium

## 1910.1027

Section 7 applies to all occupational exposures to cadmium and cadmium compounds, in all forms and in all industries, covered by the Occupational Safety and Health Act, except the construction-related industries which are covered under 29 CFR 1926.63.

## Definitions

- *Action level* (AL) is defined as an airborne concentration of cadmium of 2.5 micrograms per cubic meter of air (2.5 $mg/m^3$), calculated as an 8-hour time-weighted average (TWA).
- *Assistant Secretary* means the Assistant Secretary of Labor for Occupational Safety and Health, U.S. Department of Labor, or designee.
- *Authorized person* means any person authorized by the employer and required by work duties to be present in regulated areas or any person authorized by the OSH Act or regulations issued under it to be in regulated areas.
- *Director* means the Director of the National Institute for Occupational Safety and Health (NIOSH), U.S. Department of Health and Human Services, or designee.
- *Employee exposure* and similar language referring to the air cadmium level to which an employee is exposed means the exposure to airborne cadmium that would occur if the employee were not using respiratory protective equipment.
- *Final medical determination* is the written medical opinion of the employee's health status by the examining physician or, if multiple physician review or the alternative physician determination is invoked, it is the final, written medical finding, recommendation, or determination that emerges from that process.

- *High-efficiency particulate air* (HEPA) *filter* means a filter capable of trapping and retaining at least 99.97 percent of mono-dispersed particles of 0.3 micrometers in diameter.

- *Regulated area* means an area demarcated by the employer where an employee's exposure to airborne concentrations of cadmium exceeds, or can reasonably be expected to exceed, the permissible exposure limit (PEL).

## Permissible Exposure Limit (PEL)

- The employer shall assure that no employee is exposed to an airborne concentration of cadmium in excess of 5 micrograms per cubic meter of air (5 $mg/m^3$), calculated as an 8-hour TWA.

## Exposure Monitoring

- Each employer who has a workplace or work operation shall determine if any employee may be exposed to cadmium at or above the action level.

- Determinations of employee exposure shall be made from breathing zone air samples that reflect the monitored employee's regular, daily 8-hour TWA exposure to cadmium.

- Eight-hour TWA exposures shall be determined for each employee on the basis of one or more personal breathing zone air samples reflecting full shift exposure on each shift, for each job classification, in each work area.

- Where several employees perform the same job tasks, in the same job classification, on the same shift, in the same work area, and the length, duration, and level of cadmium exposures are similar, an employer may sample a representative fraction of the employees instead of all employees in order to meet this requirement. In representative sampling, the employer shall sample the employee(s) expected to have the highest cadmium exposures.

- The employer shall monitor employee exposures and shall base initial determinations on the monitoring results.

- Where the employer has monitored after September 14, 1991, under conditions that in all important aspects closely resemble those currently prevailing, and where that monitoring satisfies all other requirements including the accuracy and confidence levels, the employer may rely on such earlier monitoring results.

- Where the employer has objective data demonstrating that employee exposure to cadmium will not exceed the action level under the expected conditions of processing, use, or handling, the employer may rely upon such data instead of implementing initial monitoring.

- If the initial monitoring or periodic monitoring reveals employee exposures to be at or above the action level, the employer shall monitor at a frequency and pattern needed to represent the levels of exposure of employees and where exposures are above the PEL to assure the adequacy of respiratory selection and the effectiveness of engineering and work-practice controls. However, such exposure monitoring shall be performed at least every six months. The employer, at a minimum, shall continue these semi-annual measurements.

- If the initial monitoring or the periodic monitoring indicates that employee exposures are below the action level and that result is confirmed by the results of another monitoring taken at least seven days later, the employer may discontinue the monitoring for those employees whose exposures are represented by such monitoring.
- The employer also shall institute the exposure monitoring whenever there has been a change in the raw materials, equipment, personnel, work practices, or finished products that may result in additional employees being exposed to cadmium at or above the action level, or may result in employees already exposed to cadmium at or above the action level being exposed above the PEL, or whenever the employer has any reason to suspect that any other change might result in such further exposure.
- Within 15 working days after the receipt of the results of any monitoring, the employer shall notify each affected employee individually in writing of the results. In addition, within the same time period the employer shall post the results of the exposure monitoring in an appropriate location that is accessible to all affected employees.
- Wherever monitoring results indicate that employee exposure exceeds the PEL, the employer shall include in the written notice a statement that the PEL has been exceeded and a description of the corrective action being taken by the employer to reduce employee exposure to or below the PEL.
- The employer shall use a method of monitoring and analysis that has an accuracy of not less than plus or minus 25 percent, with a confidence level of 95 percent, for airborne concentrations of cadmium at or above the action level, the permissible exposure limit (PEL), and the separate engineering control air limit (SECAL).

## Regulated Areas

- The employer shall establish a regulated area wherever an employee's exposure to airborne concentrations of cadmium is, or can reasonably be expected to be, in excess of the permissible exposure limit (PEL).
- Regulated areas shall be demarcated from the rest of the workplace in any manner that adequately establishes and alerts employees of the boundaries of the regulated area.
- Access to regulated areas shall be limited to authorized persons.
- Each person entering a regulated area shall be supplied with, and required to use, a respirator.
- The employer shall assure that employees do not eat, drink, smoke, chew tobacco or gum, or apply cosmetics in regulated areas; carry the products associated with these activities into regulated areas; or store such products in those areas.

## Methods of Compliance

- ❑ The employer shall implement engineering and work-practice controls to reduce and maintain employee exposure to cadmium at or below the PEL, except to the extent that the employer can demonstrate that such controls are not feasible.

- ❑ In industries where a separate engineering control air limit (SECAL) has been specified for particular processes (See Table 4-7-1), the employer shall implement engineering and work-practice controls to reduce and maintain employee exposure at or below the SECAL, except to the extent that the employer can demonstrate that such controls are not feasible.

## Table 4-7-1: Separate Engineering Control Airborne Limits for Processes in Selected Industries

| Industry | Process | SECAL (µg/m$^3$) |
|---|---|---|
| Nickel cadmium battery | Plate making, plate preparation | 50 |
| | All other processes | 15 |
| Zinc/Cadmium refining* | Cadmium refining, casting, melting, oxide production, sinter plant | 50 |
| Pigment manufacture | Calcine, crushing, milling, blending | 50 |
| | All other processes | 15 |
| Stabilizers* | Cadmium oxide charging, crushing, drying, blending | 50 |
| Lead smelting* | Sinter plant, blast furnace, baghouse, yard area | 50 |
| Plating* | Mechanical plating | 15 |

*Processes in these industries that are not specified in this table must achieve the PEL using engineering controls and work practices.

- ❑ The requirement to implement engineering and work-practice controls to achieve the PEL or, where applicable, the SECAL, does not apply where the employer demonstrates that
  - The employee is only intermittently exposed; and
  - The employee is not exposed above the PEL on 30 or more days per year (12 consecutive months).

- ❑ Wherever engineering and work-practice controls are required, and are not sufficient to reduce employee exposure to or below the PEL, or where applicable, the SECAL, the employer nonetheless shall implement such controls to reduce exposures to the lowest levels achievable.

- ❑ The employer shall not use employee rotation as a method of compliance.

- ❑ Where the PEL is exceeded, the employer shall establish and implement a written compliance program to reduce employee exposure to or below the PEL by means of engineering and work-practice controls.

- To the extent that engineering and work-practice controls cannot reduce exposures to or below the PEL, the employer shall include in the written compliance program the use of appropriate respiratory protection to achieve compliance with the PEL.

- Written compliance programs shall include at least the following:

  - A description of each operation in which cadmium is emitted (e.g., machinery used, material processed, controls in place, crew size, employee job responsibilities, operating procedures, and maintenance practices);
  - A description of the specific means that will be employed to achieve compliance, including engineering plans and studies used to determine methods selected for controlling exposure to cadmium, as well as, where necessary, the use of appropriate respiratory protection to achieve the PEL;
  - A report of the technology considered in meeting the PEL;
  - Air monitoring data that document the sources of cadmium emissions;
  - A detailed schedule for implementation of the program, including documentation such as copies of purchase orders for equipment, construction contracts, etc.;
  - A work practice program;
  - A written plan for emergency situations; and
  - Other relevant information.

- The written compliance programs shall be reviewed and updated at least annually, or more often if necessary, to reflect significant changes in the employer's compliance status.

- Written compliance programs shall be provided upon request for examination and copying to affected employees, designated employee representatives, as well as to the Assistant Secretary, and the Director.

- When ventilation is used to control exposure, measurements that demonstrate the effectiveness of the system in controlling exposure, such as capture velocity, duct velocity, or static pressure, shall be made as necessary to maintain its effectiveness.

- Measurements of the system's effectiveness in controlling exposure shall be made as necessary within five working days of any change in production, process, or control that might result in a significant increase in employee exposure to cadmium.

- If air from exhaust ventilation is recirculated into the workplace, the system shall have a high efficiency filter and be monitored to assure effectiveness.

- Procedures shall be developed and implemented to minimize employee exposure to cadmium when maintenance of ventilation systems and changing of filters is being conducted.

## Respiratory Protection

- Respirators must be used during:
  - Periods necessary to install or implement feasible engineering and work-practice controls when employee exposure levels exceed the PEL.
  - Maintenance and repair activities, and brief or intermittent operations, for which employee exposures exceed the PEL and engineering and work-practice controls are not feasible or are not required.
  - Activities in regulated areas.
  - Work operations for which the employer has implemented all feasible engineering and work-practice controls and such controls are not sufficient to reduce employee exposures to or below the PEL.
  - Work operations for which an employee is exposed to cadmium at or above the action level, and the employee requests a respirator.
  - Work operations for which an employee is exposed to cadmium above the PEL and engineering controls are not required.
  - Emergencies.
- The employer must implement a respiratory protection program in accordance with 29 CFR 1910.134 (b) through (d) (except (d)(1)(iii)) and (f) through (m).
- No employees must use a respirator if, based on their most recent medical examination, the examining physician determines that they will be unable to continue to function normally while using a respirator.
- If an employee has breathing difficulty during fit testing or respirator use, the employer must provide the employee with a medical examination to determine if the employee can use a respirator while performing the required duties.
- The employer must select the appropriate respirator from Table 4-7-2.
- The employer must provide an employee with a powered air-purifying respirator instead of a negative-pressure respirator when an employee who is entitled to a respirator chooses to use this type of respirator and such a respirator provides adequate protection to the employee.

## Table 4-7-2: Respiratory Protection for Cadmium

| Airborne concentration or condition of use | Required respirator[a] |
|---|---|
| 10 X PEL or less | A half mask, air-purifying respirator equipped with a HEPA filter.[b] |
| 25 X PEL or less | A powered air-purifying respirator ("PAPR") with a HEPA filter, or a supplied-air respirator with a loose-fitting hood or helmet facepiece operated in the continuous flow mode. |
| 50 X PEL or less | A full-facepiece air-purifying respirator equipped with a HEPA filter, or a powered air-purifying respirator with a tight-fitting half mask equipped with a HEPA filter, or a supplied-air respirator with a tight-fitting half mask operated in the continuous flow mode. |
| 250 X PEL or less | A powered air-purifying respirator with a tight-fitting full facepiece equipped with a HEPA filter, or a supplied-air respirator with a tight-fitting full facepiece operated in the continuous flow mode. |
| 1,000 X PEL or less | A supplied-air respirator with half mask or full facepiece operated in the pressure demand or other positive-pressure mode. |
| >1,000 X PEL or unknown concentrations | A self-contained breathing apparatus with a full facepiece operated in the pressure-demand or other positive-pressure mode, or a supplied-air respirator with a full facepiece operated in the pressure-demand or other positive-pressure mode and equipped with an auxiliary escape type self-contained breathing apparatus operated in the pressure demand mode. |
| Firefighting | A self-contained breathing apparatus with full facepiece operated in the pressure-demand or other positive-pressure mode. |

[a] Respirators assigned for higher environmental concentrations may be used at lower exposure levels. Quantitative fit testing is required for all tight-fitting air-purifying respirators where airborne concentration of cadmium exceeds 10 times the TWA PEL (10 X 5 ug/m$^3$) = 50 ug/m$^3$). A full-facepiece respirator is required when eye irritation is experienced.

[b] Fit testing, qualitative or quantitative, is required.

SOURCE: Respiratory Decision Logic, NIOSH, 1987.

## Emergency Situations

- The employer shall develop and implement a written plan for dealing with emergency situations involving substantial releases of airborne cadmium.
- The plan shall include provisions for the use of appropriate respirators and personal protective equipment. In addition, employees not essential to correcting the emergency situation shall be restricted from the area, and normal operations halted in that area, until the emergency is abated.
- If an employee is exposed to airborne cadmium above the PEL, or where skin or eye irritation is associated with cadmium exposure at any level, the employer shall provide at no cost to the

employee, and assure that the employee uses, appropriate protective work clothing and equipment that prevents contamination of the employee and the employee's garments. Protective work clothing and equipment includes, but is not limited to:

- Coveralls or similar full-body work clothing;
- Gloves, head coverings, and boots or foot coverings; and
- Face shields, vented goggles, or other appropriate protective equipment that complies with 29 CFR 1910.133.

## Protective Clothing and Equipment

- ❑ The employer shall assure that employees remove all protective clothing and equipment contaminated with cadmium at the completion of the work shift and do so only in designated change rooms.

- ❑ The employer shall assure that no employee takes cadmium-contaminated protective clothing or equipment from the workplace, except for employees authorized to do so for purposes of laundering, cleaning, maintaining, or disposing of cadmium-contaminated protective clothing and equipment at an appropriate location or facility away from the workplace.

- ❑ The employer shall assure that contaminated protective clothing and equipment, when removed for laundering, cleaning, maintenance, or disposal, is placed and stored in sealed, impermeable bags or other closed, impermeable containers that are designed to prevent dispersion of cadmium dust.

- ❑ The employer shall assure that bags or containers of contaminated protective clothing and equipment that are to be taken out of the change rooms or the workplace for laundering, cleaning, maintenance, or disposal shall bear labels.

- ❑ The employer shall provide the protective clothing and equipment in a clean and dry condition as often as necessary to maintain its effectiveness, but in any event at least weekly. The employer is responsible for cleaning and laundering the protective clothing and equipment required by this paragraph to maintain its effectiveness and is also responsible for disposing of such clothing and equipment.

- ❑ The employer also is responsible for repairing or replacing required protective clothing and equipment as needed to maintain its effectiveness. When rips or tears are detected while an employee is working they shall be immediately mended, or the work suit shall be immediately replaced.

- ❑ The employer shall prohibit the removal of cadmium from protective clothing and equipment by blowing, shaking, or any other means that disperses cadmium into the air.

- ❑ The employer shall assure that any laundering of contaminated clothing or cleaning of contaminated equipment in the workplace is done in a manner that prevents the release of airborne cadmium in excess of the permissible exposure limit.

- The employer shall inform any person who launders or cleans protective clothing or equipment contaminated with cadmium of the potentially harmful effects of exposure to cadmium and that the clothing and equipment should be laundered or cleaned in a manner to effectively prevent the release of airborne cadmium in excess of the PEL.

## Hygiene Facilities and Practices

- For employees whose airborne exposure to cadmium is above the PEL, the employer shall provide clean change rooms, hand washing facilities, showers, and lunchroom facilities that comply with 29 CFR 1910.141.

- The employer shall assure that change rooms are equipped with separate storage facilities for street clothes and for protective clothing and equipment, which are designed to prevent dispersion of cadmium and contamination of the employee's street clothes.

- The employer shall assure that employees who are exposed to cadmium above the PEL shower at the end of the work shift.

- The employer shall assure that employees whose airborne exposure to cadmium is above the PEL wash their hands and faces prior to eating, drinking, smoking, chewing tobacco or gum, or applying cosmetics.

- The employer shall assure that the lunchroom facilities are readily accessible to employees, that tables for eating are maintained free of cadmium, and that no employee in a lunchroom facility is exposed at any time to cadmium at or above a concentration of 2.5 mg/m$^3$.

- The employer shall assure that employees do not enter lunchroom facilities with protective work clothing or equipment unless surface cadmium has been removed from the clothing and equipment by HEPA vacuuming or some other method that removes cadmium dust without dispersing it.

## Housekeeping

- All surfaces shall be maintained as free as practicable of accumulations of cadmium.

- All spills and sudden releases of material containing cadmium shall be cleaned up as soon as possible.

- Surfaces contaminated with cadmium shall, wherever possible, be cleaned by vacuuming or other methods that minimize the likelihood of cadmium becoming airborne.

- HEPA-filtered vacuuming equipment or equally effective filtration methods shall be used for vacuuming. The equipment shall be used and emptied in a manner that minimizes the reentry of cadmium into the workplace.

- Shoveling, dry or wet sweeping, and brushing may be used only where vacuuming or other methods that minimize the likelihood of cadmium becoming airborne have been tried and found not to be effective.

- Compressed air shall not be used to remove cadmium from any surface unless the compressed air is used in conjunction with a ventilation system designed to capture the dust cloud created by the compressed air.

- Waste, scrap, debris, bags, containers, personal protective equipment, and clothing contaminated with cadmium and consigned for disposal shall be collected and disposed of in sealed impermeable bags or other closed, impermeable containers. These bags and containers shall be labeled.

## Medical Surveillance

- The employer shall institute a medical surveillance program for all employees who are, or may be, exposed to cadmium at or above the action level unless the employer demonstrates that the employee is not, and will not be, exposed at or above the action level on 30 or more days per year (12 consecutive months).

- The employer shall also institute a medical surveillance program for all employees who might previously have been exposed to cadmium at or above the action level, prior to the effective date of this section, unless the employer demonstrates that the employee did not prior to the effective date of this section work for the employer in jobs with exposure to cadmium for an aggregated total of more than 60 months.

- To determine an employee's fitness for using a respirator, the employer shall provide the limited medical examination.

- The employer shall assure that all medical examinations and procedures are performed by or under the supervision of a licensed physician. These examinations and procedures shall be provided without cost to the employee and at a time and place that is reasonable and convenient to employees.

- The employer shall assure that the collecting and handling of biological samples of cadmium in urine (CdU), cadmium in blood (CdB), and beta-2 microglobulin in urine (ß2-M) taken from employees is done in a manner that assures their reliability and that analysis of these biological samples is performed in laboratories with demonstrated proficiency for that particular analyte.

- The employer shall provide an initial (pre-placement) examination to all employees covered by the medical surveillance program. The examination shall be provided to those employees within 30 days after initial assignment to a job with exposure to cadmium.

- The initial (pre-placement) medical examination shall include:

  - A detailed medical and work history with emphasis on past, present, and anticipated future exposure to cadmium; any history of renal, cardiovascular, respiratory, hematopoietic, reproductive, and/or musculo-skeletal system dysfunction; current usage of medication with potential nephrotoxic side-effects; and smoking history and current status; and
  - Biological monitoring that includes the following tests:
    - Cadmium in urine (CdU), standardized to grams of creatinine (g/Cr);

   - Beta-2 microglobulin in urine (ß2-M), standardized to grams of creatinine (g/Cr), with pH specified; and
   - Cadmium in blood (CdB), standardized to liters of whole blood (lwb).

- An initial examination is not required if adequate records show that the employee has been examined within the past 12 months. In that case, such records shall be maintained as part of the employee's medical record and the prior exam shall be treated as if it were an initial examination.

- If the results of the initial biological monitoring tests show the employee's CdU level to be at or below 3 mg/g Cr, ß2-M level to be at or below 300 mg/g Cr, and CdB level to be at or below 5 mg/lwb, then:
   - For currently exposed employees who are subject to medical surveillance, the employer shall provide the minimum level of periodic medical surveillance.
   - For previously exposed employees who are subject to medical surveillance, the employer shall provide biological monitoring for CdU, ß2-M, and CdB one year after the initial biological monitoring.

- For all employees who are subject to medical surveillance, if the results of the initial biological monitoring tests show the level of CdU to exceed 3 mg/g Cr, the level of ß2-M to exceed 300 mg/g Cr, or the level of CdB to exceed 5 mg/lwb, the employer shall:
   - Within two weeks after receipt of biological monitoring results, reassess the employee's occupational exposure to cadmium as follows:
      - Assess the employee's work practices and personal hygiene;
      - Evaluate the employee's respirator use, if any, and the respirator program;
      - Review the hygiene facilities;
      - Evaluate the maintenance and effectiveness of the relevant engineering controls; and
      - Assess the employee's smoking history and status.
   - Within 30 days after the exposure reassessment, take reasonable steps to correct any deficiencies found in the reassessment that may be responsible for the employee's excess exposure to cadmium; and
   - Within 90 days after receipt of biological monitoring results, provide a full medical examination to the employee.

- After completing the medical examination, the examining physician shall determine in a written medical opinion whether to medically remove the employee. If the physician determines that medical removal is not necessary, then until the employee's CdU level falls to or below 3 mg/g Cr, ß2-M level falls to or below 300 mg/g Cr, and CdB level falls to or below 5 mg/lwb, the employer shall:
   - Provide biological monitoring on a semiannual basis; and
   - Provide annual medical examinations.

- For all employees who are subject to medical surveillance, within 90 days after receipt of biological monitoring results, the employer shall provide a full medical examination to the employee.

- After completing the medical examination, the examining physician shall determine in a written medical opinion whether to medically remove the employee. However, the physician shall medically remove the employee from exposure to cadmium at or above the action level, if the initial biological monitoring results and the biological monitoring results obtained during the medical examination both show that the CdU level exceeds 15 mg/g Cr; the CdB level exceeds 15 mg/lwb; or the ß2-M level exceeds 1,500 mg/g Cr, and in addition the CdU exceeds 3 mg/g Cr, or CdB level exceeds 5 mg/liter of whole blood.

- If the second set of biological monitoring results obtained during the medical examination does not show that a mandatory removal trigger level has been exceeded, then the employee is not required to be removed by the mandatory provisions of this paragraph. If the employee is not required to be removed by the mandatory provisions of this paragraph or by the physician's determination, then until the employee's CdU level falls to or below 3 mg/g Cr, ß2-M level falls to or below 300 mg/g Cr, and CdB level falls to or below 5 mg/lwb, the employer shall:
  - Periodically reassess the employee's occupational exposure to cadmium;
  - Provide biological monitoring on a quarterly basis; and
  - Provide semi-annual medical examinations.

- For all employees to whom medical surveillance is provided, beginning on January 1, 1999:
  - The employer shall within 90 days after receipt of biological monitoring results, provide a full medical examination to the employee:
    - If the results of the initial biological monitoring tests show the employee's CdU level to be at or below 3 mg/g Cr, ß2-M level to be at or below 300 mg/g Cr, and CdB level to be at or below 5 mg/lwb.
    - If the results of the initial biological monitoring tests show the level of CdU to exceed 3 mg/g Cr, the level of ß2-M to exceed 300 mg/g Cr, or the level of CdB to exceed 5 mg/lwb.
    - If the results of the initial biological monitoring tests show the level of CdU to be in excess of 7 mg/g Cr, or the level of CdB to be in excess of 10 mg/lwb, or the level of ß2-M to be in excess of 750 mg/g Cr.
  - After completing the medical examination, the examining physician shall determine in a written medical opinion whether to medically remove the employee.
  - The physician shall medically remove the employee from exposure to cadmium at or above the action level, if the initial biological monitoring results and the biological monitoring results obtained during the medical examination both show that the CdU level exceeds 7 mg/g Cr, the CdB level exceeds 10 mg/lwb, the ß2-M level exceeds 750 mg/g Cr, and in addition the CdU level exceeds 3 mg/g Cr or the CdB level exceeds 5 mg/liter of whole blood.

- If the second set of biological monitoring results obtained during the medical examination does not show that a mandatory removal trigger level has been exceeded, then the employee is not required to be removed.
- If the employee is not required to be removed by the physician's determination, then until the employee's CdU level falls to or below 3 mg/g Cr, ß2-M level falls to or below 300 mg/g Cr, and CdB level falls to or below 5 mg/lwb, the employer shall:
    - Periodically reassess the employee's occupational exposure to cadmium;
    - Provide biological monitoring on a quarterly basis; and
    - Provide semi-annual medical examinations.

- ❑ The employer shall provide at least the minimum level of periodic medical surveillance, which consists of periodic medical examinations and periodic biological monitoring.

- ❑ A periodic medical examination shall be provided within one year after the initial examination and thereafter at least biennially.

- ❑ Biological sampling shall be provided at least annually, either as part of a periodic medical examination or separately as periodic biological monitoring.

- ❑ The periodic medical examination shall include:

    - A detailed medical and work history, or update thereof, with emphasis on past, present, and anticipated future exposure to cadmium; smoking history and current status; reproductive history; current use of medications with potential nephrotoxic side-effects; any history of renal, cardiovascular, respiratory, hematopoietic, and/or musculo-skeletal system dysfunction;
    - A complete physical examination with emphasis on blood pressure, the respiratory system, and the urinary system;
    - A 14 x 17 inch, or a reasonably standard sized, posterior-anterior chest X-ray (after the initial X-ray, the frequency of chest X-rays is to be determined by the examining physician);
    - Pulmonary function tests, including forced vital capacity (FVC) and forced expiratory volume at 1 second (FEV1);
    - Biological monitoring;
    - Blood analysis, including blood urea nitrogen, complete blood count, and serum creatinine;
    - Urinalysis, including the determination of albumin, glucose, and total and low molecular weight proteins;
    - For males over 40 years old, prostate palpation, or other at least as effective diagnostic test(s); and
    - Any additional tests deemed appropriate by the examining physician.

## Periodic Biological Monitoring

- ❑ The employer shall take appropriate actions if the results of periodic biological monitoring or the results of biological monitoring performed as part of the periodic medical examination show elevated levels of the employee's CdU, ß2-M, or CdB levels.

- ❑ For previously exposed employees:

  - If the employee's levels of CdU did not exceed 3 mg/g Cr, CdB levels did not exceed 5 mg/lwb, and ß2-M levels did not exceed 300 mg/g Cr in the initial biological monitoring tests, and if the results of the follow-up biological monitoring one year after the initial examination confirm the previous results, the employer may discontinue all periodic medical surveillance for that employee.
  - If the initial biological monitoring results for CdU, CdB, or ß2-M were in excess of the levels, but subsequent biological monitoring results show that the employee's CdU level no longer exceeds 3 mg/g Cr, CdB level no longer exceeds 5 mg/lwb, and ß2-M level no longer exceeds 300 mg/g Cr, the employer shall provide biological monitoring for CdU, CdB, and ß2-M one year after these most recent biological monitoring results.
  - If the results of the follow-up biological monitoring specified in this paragraph confirm the previous results, the employer may discontinue all periodic medical surveillance for that employee.
  - If the results of the follow-up tests indicate that the level of the employee's CdU, ß2-M, or CdB exceeds these same levels, the employer is required to provide annual medical examinations until the results of biological monitoring are consistently below these levels or the examining physician determines in a written medical opinion that further medical surveillance is not required to protect the employee's health.

- ❑ A routine, biennial medical examination is not required if adequate medical records show that the employee has been examined within the past 12 months.

- ❑ Such records shall be maintained by the employer as part of the employee's medical record, and the next routine, periodic medical examination shall be made available to the employee within two years of the previous examination.

- ❑ If the results of a medical examination carried out in accordance with this section indicate any laboratory or clinical finding consistent with cadmium toxicity that does not require employer action, the employer, within 30 days, shall reassess the employee's occupational exposure to cadmium and take the following corrective actions until the physician determines they are no longer necessary:

  - Periodically reassess the employee's work practices and personal hygiene; the employee's respirator use, if any; the employee's smoking history and status; the respiratory protection program; the hygiene facilities; and the maintenance and effectiveness of the relevant engineering controls;
  - Within 30 days after the reassessment, take all reasonable steps to correct the deficiencies found in the reassessment that may be responsible for the employee's excess exposure to

cadmium;

- Provide semi-annual medical reexaminations to evaluate the abnormal clinical signs of cadmium toxicity until the results are normal or the employee is medically removed; and
- Where the results of tests for total proteins in urine are abnormal, provide a more detailed medical evaluation of the toxic effects of cadmium on the employee's renal system.

- To determine an employee's fitness for respirator use, the employer shall provide a medical examination that includes:
  - A detailed medical and work history, or update thereof, with emphasis on past exposure to cadmium; smoking history and current status; any history of renal, cardiovascular, respiratory, hematopoietic, and/or musculoskeletal system dysfunction; and a description of the job for which the respirator is required;
  - A blood pressure test;
  - Biological monitoring of the employee's levels of CdU, CdB, and ß2-M, unless such results already have been obtained within the previous 12 months; and
  - Any other test or procedure that the examining physician deems appropriate.
- After reviewing all the information obtained from the medical examination, the physician shall determine whether the employee is fit to wear a respirator.
- Whenever an employee has exhibited difficulty in breathing during a respirator fit test or use of a respirator, the employer, as soon as possible, shall provide the employee with a medical examination to determine the employee's fitness to wear a respirator.
- Where the results of the examination are abnormal, medical limitation or prohibition of respirator use shall be considered. If the employee is allowed to wear a respirator, the employee's ability to continue to do so shall be periodically evaluated by a physician.
- In addition to the medical surveillance, the employer shall provide a medical examination as soon as possible to any employee who may have been acutely exposed to cadmium because of an emergency.
- The examination shall include, with emphasis on the respiratory system, other organ systems considered appropriate by the examining physician, and symptoms of acute overexposure.
- At termination of employment, the employer shall provide a medical examination, including a chest X-ray, to any employee to whom at any prior time the employer was required to provide medical surveillance. However, if the last examination was less than six months prior to the date of termination, no further examination is required.
- If the employer has discontinued all periodic medical surveillance, no termination of employment medical examination is required.
- The employer shall provide the following information to the examining physician:
  - A copy of this standard and appendices;

- A description of the affected employee's former, current, and anticipated duties as they relate to the employee's occupational exposure to cadmium;
- The employee's former, current, and anticipated future levels of occupational exposure to cadmium;
- A description of any personal protective equipment, including respirators, used or to be used by the employee, including when and for how long the employee has used the equipment; and
- Relevant results of previous biological monitoring and medical examinations.

❑ The employer shall promptly obtain a written, signed medical opinion from the examining physician for each medical examination performed on each employee. This written opinion shall contain:

- The physician's diagnosis for the employee;
- The physician's opinion as to whether the employee has any detected medical condition(s) that would place the employee at increased risk of material impairment to health from further exposure to cadmium, including any indications of potential cadmium toxicity;
- The results of any biological or other testing or related evaluations that directly assess the employee's absorption of cadmium;
- Any recommended removal from, or limitation on the activities or duties of the employee or on the employee's use of personal protective equipment, such as respirators; and
- A statement that the physician has clearly and carefully explained to the employee the results of the medical examination, including all biological monitoring results, and any medical conditions related to cadmium exposure that require further evaluation or treatment, and any limitation on the employee's diet or use of medications.

❑ The employer promptly shall obtain a copy of the results of any biological monitoring provided by an employer to an employee independently of a medical examination, and, in lieu of a written medical opinion, an explanation sheet explaining those results.

❑ The employer shall instruct the physician not to reveal orally or in the written medical opinion given to the employer specific findings or diagnoses unrelated to occupational exposure to cadmium.

❑ The employer shall temporarily remove an employee from work where there is excess exposure to cadmium on each occasion that medical removal is required and on each occasion that a physician determines in a written medical opinion that the employee should be removed from such exposure. The physician's determination may be based on biological monitoring results, inability to wear a respirator, evidence of illness, other signs or symptoms of cadmium-related dysfunction or disease, or any other reason deemed medically sufficient by the physician.

❑ The employer shall medically remove an employee regardless of whether at the time of removal a job is available into which the removed employee may be transferred.

❑ Whenever an employee is medically removed, the employer shall transfer the removed employee to a job where the exposure to cadmium is within the permissible levels as soon as one becomes available.

- For any employee who is medically removed, the employer shall provide follow-up biological monitoring at least every three months and follow-up medical examinations semi-annually at least every six months until the examining physician determines in a written medical opinion tthat either the employee may be returned to his/her former job status or the employee must be permanently removed from excess cadmium exposure.

- The employer may not return an employee who has been medically removed for any reason to his/her former job status until a physician determines in a written medical opinion that continued medical removal is no longer necessary to protect the employee's health.

- Where an employee is found unfit to wear a respirator, the employer shall remove the employee from work where exposure to cadmium is above the PEL.

- Where removal is based on any reason other than the employee's inability to wear a respirator, the employer shall remove the employee from work where exposure to cadmium is at or above the action level.

- No employee who was removed because his/her level of CdU, CdB, and/or ß2-M exceeded the medical removal trigger levels may be returned to work with exposure to cadmium at or above the action level until the employee's level of CdU falls to or below 3 mg/g Cr, CdB level falls to or below 5 mg/lwb, and ß2-M level falls to or below 300 mg/g Cr.

- However, when the examining physician's opinion is that continued exposure to cadmium will not pose an increased risk to the employee's health and there are special circumstances that make continued medical removal an inappropriate remedy, the physician shall fully discuss these matters with the employee, and then in a written determination may return a worker to his/her former job status despite what would otherwise be unacceptably high biological monitoring results. Thereafter, the returned employee shall continue to be provided with medical surveillance as if he/she were still on medical removal until the employee's level of CdU falls to or below 3 mg/g Cr, CdB level falls to or below 5 mg/lwb, and ß2-M level falls to or below 300 mg/g Cr.

- Where an employer removes an employee from exposure to cadmium or otherwise places limitations on an employee due to the effects of cadmium exposure on the employee's medical condition, the employer shall provide the same medical removal protection benefits (MRPB) to that employee.

- The employer shall provide MRPB for up to a maximum of 18 months to an employee each time, and while the employee is temporarily medically removed.

- The requirement that the employer provide MRPB means that the employer shall maintain the total normal earnings, seniority, and all other employee rights and benefits of the removed employee, including the employee's right to his/her former job status, as if the employee had not been removed from the employee's job or otherwise medically limited.

- Where, after 18 months on medical removal because of elevated biological monitoring results, the employee's monitoring results have not declined to a low enough level to permit the employee to be returned to his/her former job status:

- The employer shall make available to the employee a medical examination in order to obtain a final medical determination as to whether the employee may be returned to his/her former job status or must be permanently removed from excess cadmium exposure.
- The employer shall assure that the final medical determination indicates whether the employee may be returned to his/her former job status and what steps, if any, should be taken to protect the employee's health.

❑ The employer may condition the provision of MRPB upon the employee's participation in medical surveillance provided in accordance with this section.

❑ If the employer selects the initial physician to conduct any medical examination or consultation provided to an employee under this section, the employee may designate a second physician to:

- Review any findings, determinations, or recommendations of the initial physician; and
- Conduct such examinations, consultations, and laboratory tests as the second physician deems necessary to facilitate this review.

❑ The employer shall promptly notify an employee of the right to seek a second medical opinion after each occasion that an initial physician provided by the employer conducts a medical examination or consultation pursuant to this section. The employer may condition its participation in, and payment for, multiple physician reviews upon the employee doing the following within 15 days after receipt of this notice, or receipt of the initial physician's written opinion, whichever is later:

- Informing the employer that he or she intends to seek a medical opinion; and
- Initiating steps to make an appointment with a second physician.

❑ If the findings, determinations, or recommendations of the second physician differ from those of the initial physician, then the employer and the employee shall assure that efforts are made for the two physicians to resolve any disagreement.

❑ If the two physicians have been unable to quickly resolve their disagreement, then the employer and the employee, through their respective physicians, shall designate a third physician to:

- Review any findings, determinations, or recommendations of the other two physicians; and
- Conduct such examinations, consultations, laboratory tests, and discussions with the other two physicians as the third physician deems necessary to resolve the disagreement among them.

❑ The employer shall act consistently with the findings, determinations, and recommendations of the third physician, unless the employer and the employee reach an agreement that is consistent with the recommendations of at least one of the other two physicians.

❑ The employer and employee, or designated employee representative, may agree upon the use of any alternate form of physician determination in lieu of the multiple physician review, as long as the alternative is expeditious and at least as protective of the employee.

❑ The employer shall provide a copy of the physician's written medical opinion to the examined employee within two weeks after its receipt.

- The employer shall provide the employee with a copy of the employee's biological monitoring results and an explanation sheet explaining the results within two weeks after its receipt.
- Within 30 days after a request by an employee, the employer shall provide the employee with the information the employer is required to provide the examining physician.
- In addition to other medical events that are required to be reported on the OSHA Form No. 200, the employer shall report any abnormal condition or disorder caused by occupational exposure to cadmium associated with employment as specified in Chapter (V)(E) of the Reporting Guidelines for Occupational Injuries and Illnesses.

## Communication of Hazards to Employees

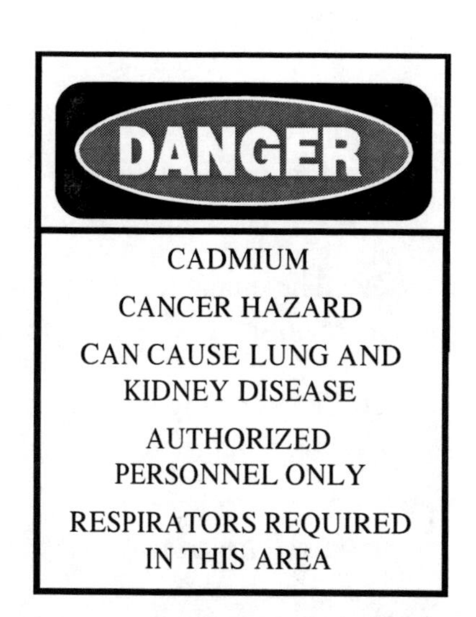

- In communications concerning cadmium hazards, employers shall comply with the requirements of OSHA's Hazard Communication Standard, 29 CFR 1910.1200, including but not limited to the requirements concerning warning signs and labels, material safety data sheets (MSDS), and employee information and training.
- Warning signs shall be provided and displayed in regulated areas. In addition, warning signs shall be posted at all approaches to regulated areas so that any employee may read the signs and take necessary protective steps before entering the area. (See sample at above right.)
- The employer shall assure that signs are illuminated, cleaned, and maintained as necessary so that the legend is readily visible.
- Shipping and storage containers containing cadmium, cadmium compounds, or cadmium contaminated clothing, equipment, waste, scrap, or debris shall bear appropriate warning labels. (See sample at below right.)
- Where feasible, installed cadmium products shall have a visible label or other indication that cadmium is present.
- The employer shall institute a training program for all employees who are potentially exposed to cadmium, assure employee participation in the program, and maintain a record of the contents of such program.
- Training shall be provided prior to, or at the time of, initial assignment to a job involving potential exposure to cadmium and at least annually thereafter.
- The employer shall make the training program understandable to the employee and shall assure that each employee is informed of the following:

- The health hazards associated with cadmium exposure;
- The quantity, location, manner of use, release, and storage of cadmium in the workplace and the specific nature of operations that could result in exposure to cadmium, especially exposures above the PEL;
- The engineering controls and work practices associated with the employee's job assignment;
- The measures employees can take to protect themselves from exposure to cadmium, including modification of such habits as smoking and personal hygiene, and specific procedures the employer has implemented to protect employees from exposure to cadmium such as appropriate work practices, emergency procedures, and the provision of personal protective equipment;
- The purpose, proper selection, fitting, proper use, and limitations of respirators and protective clothing;
- The purpose and a description of the medical surveillance program;
- The contents of 29 CFR 1910.1027; and
- The employee's rights of access to records under 29 CFR 1910.20(e) and (g).

❑ The employer shall make a copy of 29 CFR 1910.1027 and its appendices readily available without cost to all affected employees and shall provide a copy if requested.

❑ The employer shall provide, upon request, to the Assistant Secretary or the Director all materials relating to the employee information and training program.

## Recordkeeping

❑ The employer shall establish and keep an accurate record of all air monitoring for cadmium in the workplace.

❑ This record shall include at least the following information:

- The monitoring date, duration, and results in terms of an 8-hour TWA of each sample taken;
- The name, social security number, and job classification of the employees monitored and of all other employees whose exposures the monitoring is intended to represent;
- A description of the sampling and analytical methods used, and evidence of their accuracy;
- The type of respiratory protective device, if any, worn by the monitored employee; and
- A notation of any other conditions that might have affected the monitoring results.

❑ The employer shall maintain this record for at least 30 years, in accordance with 29 CFR 1910.20.

❑ Objective data are information demonstrating that a particular product or material containing cadmium or a specific process, operation, or activity involving cadmium cannot release dust or fumes in concentrations at or above the action level even under the worst-case release conditions.

- Objective data can be obtained from an industry-wide study or from laboratory product test results from manufacturers of cadmium-containing products or materials.

- The data the employer uses from an industry-wide survey must be obtained under workplace conditions closely resembling the processes, types of material, control methods, work practices and environmental conditions in the employer's current operations.

- The employer shall establish and maintain a record of the objective data for at least 30 years.

- The employer shall establish and maintain an accurate record for each employee covered by medical surveillance.

- The record shall include at least the following information about the employee:
  - Name, social security number, and description of the duties;
  - A copy of the physician's written opinions and an explanation sheet for biological monitoring results;
  - A copy of the medical history, and the results of any physical examination and all test results that are required to be provided by this section, including biological tests, X-rays, pulmonary function tests, etc., or that have been obtained to further evaluate any condition that might be related to cadmium exposure;
  - The employee's medical symptoms that might be related to exposure to cadmium; and
  - A copy of the information provided to the physician.

- The employer shall assure that this record is maintained for the duration of employment plus 30 years, in accordance with 29 CFR 1910.20.

- The employer shall certify that employees have been trained by preparing a certification record which includes the identity of the person trained, the signature of the employer or the person who conducted the training, and the date the training was completed. The certification records shall be prepared at the completion of training and shall be maintained on file for one year beyond the date of training of that employee.

- Except as otherwise provided for in this section, access to all records shall be in accordance with the provisions of 29 CFR 1910.20.

- Within 15 days after a request, the employer shall make an employee's medical records available for examination and copying to the subject employee, to designated representatives, to anyone having the specific written consent of the subject employee, and, after the employee's death or incapacitation, to the employee's family members.

- Whenever an employer ceases to do business and there is no successor employer to receive and retain records for the prescribed period, or the employer intends to dispose of any records required to be preserved for at least 30 years, the employer shall comply with the requirements concerning transfer of records set forth in 29 CFR 1910.20(h).

## Observation of Monitoring

- ❑ The employer shall provide affected employees or their designated representatives an opportunity to observe any monitoring of employee exposure to cadmium.

- ❑ When observation of monitoring requires entry into an area where the use of protective clothing or equipment is required, the employer shall provide the observer with that clothing and equipment and shall assure that the observer uses such clothing and equipment and complies with all other applicable safety and health procedures.

# Cotton Dust

## 1910.1043

Section 8 applies to the control of employee exposure to cotton dust in all workplaces where employees engage in yarn manufacturing, slashing and weaving operations, or work in waste houses for textile operations. This section does not apply to the handling or processing of woven or knitted materials; maritime operations covered by 29 CFR Parts 1915 and 1918; harvesting or ginning of cotton; or the construction industry.

### Definitions

- *Assistant Secretary* means the Assistant Secretary of Labor for Occupational Safety and Health, U.S. Department of Labor, or designee.

- *Blow down* means the general cleaning of a room or a part of a room by the use of compressed air.

- *Blow off* means the use of compressed air for cleaning of short duration and usually for a specific machine or portion of a machine.

- *Cotton dust* means dust present in the air during the handling or processing of cotton, which may contain a mixture of many substances including ground up plant matter, fiber, bacteria, fungi, soil, pesticides, non-cotton plant matter, and other contaminants which may have accumulated with the cotton during the growing, harvesting, and subsequent processing or storage periods. Any dust present during the handling and processing of cotton through the weaving or knitting of fabrics, and dust present in other operations or manufacturing processes using raw or waste cotton fibers or cotton fiber byproducts from textile mills are considered cotton dust within this definition. Lubricating oil mist associated with weaving operations is not considered cotton dust.

- *Director* means the Director of the National Institute for Occupational Safety and Health (NIOSH), U.S. Department of Health and Human Services, or designee.

- *Equivalent instrument* means a cotton dust sampling device that meets the vertical elutriator equivalency requirements.

- *Lint-free respirable cotton dust* means particles of cotton dust of approximately 15 micrometers or less aerodynamic equivalent diameter.

- *Vertical elutriator cotton dust sampler* or *vertical elutriator* means a dust sampler which has a particle size cut-off at approximately 15 micrometers or aerodynamic equivalent diameter when operating at the flow rate of $7.4 \pm 0.2$ liters of air per minute.

- *Waste processing* means waste recycling (sorting, blending, cleaning, and willowing) and garnetting.

- *Yarn manufacturing* means all textile mill operations from opening to, but not including, slashing and weaving.

## Permissible Exposure Limits (PELs) and Action Levels

- The employer shall assure that no employee who is exposed to cotton dust in yarn manufacturing and cotton washing operations is exposed to airborne concentrations of lint-free respirable cotton dust greater than 200 mg/m$^3$ mean concentration, averaged over an 8-hour period, as measured be a vertical elutriator or an equivalent instrument.

- The employer shall assure that no employee who is exposed to cotton dust in textile mill waste-house operations or who is exposed in yarn manufacturing to dust from "lower grade washed cotton" is exposed to airborne concentrations of lint-free respirable cotton dust greater than 500 mg/m$^3$ mean concentration, averaged over an 8-hour period, as measured by a vertical elutriator or an equivalent instrument.

- The employer shall assure that no employee who is exposed to cotton dust in the textile processes known as slashing and weaving is exposed to airborne concentrations of lint-free respirable cotton dust greater than 750 mg/m$^3$ mean concentration, averaged over an 8-hour period, as measured by a vertical elutriator or an equivalent instrument.

- The action level for yarn manufacturing and cotton washing operations is an airborne concentration of lint-free respirable cotton dust of 100 mg/m$^3$ mean concentration, averaged over an 8-hour period, as measured by a vertical elutriator or an equivalent instrument.

- The action level for waste houses for textile operations is an airborne concentration of lint-free respirable cotton dust of 250 mg/m$^3$ mean concentration, averaged over an 8-hour period, as measured by a vertical elutriator or an equivalent instrument.

- The action level for the textile processes known as slashing and weaving is an airborne concentration of lint-free respirable cotton dust of 375 mg/m$^3$ mean concentration, averaged over an 8-hour period, as measured by a vertical elutriator or an equivalent instrument.

## Exposure Monitoring and Measurement

- For the purposes of this section, employee exposure is that exposure which would occur if the employee were not using a respirator.

- The sampling device to be used shall be either the vertical elutriator cotton dust sampler or an equivalent instrument.

- If an alternative to the vertical elutriator cotton dust sampler is used, the employer shall establish equivalency by reference to an OSHA opinion or by documenting, based on data developed by the employer or supplied by the manufacturer, that the alternative sampling devices meets the following criteria:

  - Collect respirable particulates in the same range as the vertical elutriator (approximately 15 microns);
  - Collect replicate exposure data used to establish equivalency in side-by-side field and laboratory comparisons; and
  - Collect a minimum of 100 samples over the range of 0.5 to 2 times the permissible exposure limit, and 90 percent of these samples have an accuracy range of plus or minus 25 percent of the vertical elutriator reading with a 95 percent confidence level as demonstrated by a statistically valid protocol.

- OSHA will issue a written opinion stating that an instrument is equivalent to a vertical elutriator cotton dust sampler if:

  - A manufacturer or employer requests an opinion in writing and supplies the following information:
    - Sufficient test data to demonstrate that the instrument meets requirements;
    - Any other relevant information about the instrument and its testing requested by OSHA; and
    - A certification by the manufacturer or employer that the information supplied is accurate.
  - OSHA finds, based on information submitted about the instrument, that the instrument meets the requirements.

- Each employer shall conduct monitoring by obtaining measurements which are representative of the exposure of all employees to airborne concentrations of lint-free respirable cotton dust over an 8-hour period. The sampling program shall include at least one determination during each shift for each work area.

- If the initial monitoring or any subsequent monitoring reveals employee exposure to be at or below the PEL, the employer shall repeat the monitoring for those employees at least annually.

- If the initial monitoring or any subsequent monitoring reveals employee exposure to be above the PEL, the employer shall repeat the monitoring for those employees at least every six months.

- ❑ Whenever there has been a production, process, or control change which may result in new or additional exposure to cotton dust, or the employer has any other reason to suspect an increase in employee exposure, the employer shall repeat the monitoring and measurements for those employees affected by the change or increase.

- ❑ Within 20 working days after the receipt of monitoring results, the employer shall notify each employee in writing of the exposure measurements which represent that employee's exposure.

- ❑ Whenever the results indicate that the employee's exposure exceeds the applicable PEL, the employer shall include in the written notice a statement that the PEL was exceeded and a description of the corrective action taken to reduce exposure below the PEL .

## Methods of Compliance

- ❑ The employer shall institute engineering and work-practice controls to reduce and maintain employee exposure to cotton dust at or below the permissible exposure limit, except to the extent that the employer can establish that such controls are not feasible.

- ❑ Whenever feasible engineering and work-practice controls are not sufficient to reduce employee exposure to or below the PEL, the employer shall institute these controls to reduce exposure to the lowest feasible level, and shall supplement these controls with the use of respirators.

- ❑ Where the most recent exposure monitoring data indicates that any employee is exposed to cotton dust levels greater than the PEL, the employer shall establish and implement a written program sufficient to reduce exposures to or below the PEL solely by means of engineering controls and work practices.

- ❑ The written program shall include at least the following:
  - A description of each operation or process resulting in employee exposure to cotton dust at levels greater than the PEL;
  - Engineering plans and other studies used to determine the controls for each process;
  - A report of the technology considered in meeting the PEL;
  - Monitoring data obtained;
  - A detailed schedule for development and implementation of engineering and work-practice controls, including exposure levels projected to be achieved by such controls;
  - Work-practice program; and
  - Other relevant information.

- ❑ Written programs shall be submitted, upon request, to the Assistant Secretary and the Director, and shall be available at the worksite for examination and copying by the Assistant Secretary, the Director, and any affected employees or their designated representatives.

- ❑ The written program shall be revised and updated when necessary to reflect the current status of the program and current exposure levels.

- When mechanical ventilation is used to control exposure, measurements which demonstrate the effectiveness of the system to control exposure, such as capture velocity, duct velocity, or static pressure, shall be made at reasonable intervals.

## Respiratory Protection

- For employees who are required to use respirators, the employer must provide respirators during:
  - Periods necessary to install or implement feasible engineering and work-practice controls.
  - Maintenance and repair activities for which engineering and work-practice controls are not feasible.
  - Work operations for which feasible engineering and work-practice controls are not yet sufficient to reduce employee exposure to or below the PEL.
  - Periods for which an employee requests a respirator.
- The employer must implement a respiratory protection program in accordance with 29 CFR 1910.134(b) through (d) (except (d)(1)(iii)) and (f) through (m).
- Whenever a physician determines that an employee who works in an area in which the cotton-dust concentration exceeds the PEL is unable to use a respirator, including a powered air-purifying respirator, the employee must be given the opportunity to transfer to an available position, or to a position that becomes available later, that has a cotton-dust concentration at or below the PEL. The employer must ensure that such employees retain their current wage rate or other benefits as a result of the transfer.
- The employer must select the appropriate respirator from Table 4-8-1.
- Whenever respirators are required by this section for cotton-dust concentrations that do not exceed the applicable PEL by a multiple of 100 (100X), the employer must, when requested by an employee, provide a powered air-purifying respirator with a high-efficiency particulate filter instead of the respirator specified in Table 4-8-1.

## Table 4-8-1: Respiratory Protection for Cotton Dust

| Cotton dust concentration | Required respirator |
|---|---|
| 5 X the applicable (PEL) or less | A disposable respirator with a particulate filter. |
| 10 X the applicable PEL or less | A quarter or half-mask respirator, other than a disposable respirator, equipped with particulate filters. |
| 100 X the applicable PEL or less | A full facepiece respirator equipped with high-efficiency particulate filters. |
| Greater than 100 X the applicable PEL | A powered air-purifying respirator equipped with high-efficiency particulate filters. |

Notes:
1. A disposable respirator means the filter element is an inseparable part of the respirator.
2. Any respirators permitted at higher environmental concentrations can be used at lower concentrations.
3. Self-contained breathing apparatus are not required respirators, but are permitted respirators.
4. Supplied-air respirators are not required, but are permitted under the following conditions: Cotton dust concentration not greater than 10 X the PEL—Any supplied-air respirator; not greater than 100 X the PEL—Any supplied-air respirator with full facepiece, helmet, hood; or greater than 100 X the PEL—A supplied-air respirator operated in positive pressure mode.

## Written Program

- ❑ Each employer shall, regardless of the level of employee exposure, immediately establish and implement a written program of work practices that shall minimize cotton dust exposure. The following shall be included where applicable:
  - Compressed air "blow down" cleaning shall be prohibited where alternative means are feasible. Where compressed air is used for cleaning, the employees performing the "blow down" or "blow off" shall wear suitable respirators. Employees whose presence is not required to perform "blow down" or "blow off" shall be required to leave the area affected by the "blow down" or "blow off" during this cleaning operation.
  - Cleaning of clothing or floors with compressed air shall be prohibited.
  - Floor sweeping shall be performed with a vacuum or with methods designed to minimize dispersal of dust.
  - In areas where employees are exposed to concentrations of cotton dust greater than the PEL, cotton and cotton waste shall be stacked, sorted, baled, dumped, removed, or otherwise handled by mechanical means, except where the employer can show that it is infeasible to do so. Where infeasible, the method used for handling cotton and cotton waste shall be the method that reduces exposure to the lowest level feasible.

# Medical Surveillance

- Each employer covered by the standard shall institute a program of medical surveillance for all employees exposed to cotton dust.
- The employer shall assure that all medical examinations and procedures are performed by or under the supervision of a licensed physician and are provided without cost to the employee.
- Persons other than licensed physicians, who administer the pulmonary function testing required by this section, shall have completed a NIOSH-approved training course in spirometry.
- The employer shall provide medical surveillance to each employee who is, or may be, exposed to cotton dust. For new employees, this examination shall be provided prior to initial assignment. The medical surveillance shall include at least the following:
  - A medical history;
  - The standardized questionnaire contained in 29 CFR 1910.1043 Appendix B; and
  - A pulmonary function measurement, including a determination of forced vital capacity (FVC), forced expiratory volume in one second (FEV1), the FEV1/FVC ratio, and the percentage that the measured values of FEV1 and FVC differ from the predicted values, using the standard tables in 29 CFR 1910.1043 Appendix C. These determinations shall be made for each employee before the employee enters the workplace on the first day of the work week, preceded by at least 35 hours of no exposure to cotton dust. The tests shall be repeated during the shift, no less than four and no more than ten hours after the beginning of the work shift; and, in any event, no more than one hour after cessation of exposure. Such exposure shall be typical of the employee's usual workplace exposure. The predicted FEV1 and FVC for blacks shall be multiplied by 0.85 to adjust for ethnic differences.
- Based upon the questionnaire results, each employee shall be graded according to Schilling's byssinosis classification system.
- The employer shall provide at least annual medical surveillance for all employees exposed to cotton dust above the action level in yarn manufacturing, slashing and weaving, cotton washing, and waste-house operations. The employer shall provide medical surveillance at least every two years for all employees exposed to cotton dust at or below the action level, all employees exposed to cotton dust from washed cotton, and all employees exposed to cotton dust in cottonseed processing and waste processing operations.
- Medical surveillance shall be provided every six months for all employees who meet one of the following criteria:
  - Have an FEV1 of greater than 80 percent of the predicted value, but with an FEV1 decrement of 5 percent or 200 ml. on a first working day.
  - Have an FEV1 of less than 80 percent of the predicted value.
  - Where, in the opinion of the physician, any significant change in questionnaire findings, pulmonary function results, or other diagnostic tests have occurred.

- ❑ An employee whose FEV1 is less than 60 percent of the predicted value shall be referred to a physician for a detailed pulmonary examination.

- ❑ A comparison shall be made between the current examination results and those of previous examination, and a determination made by the physician as to whether there has been a significant change.

- ❑ The employer shall provide the following information to the examining physician:
  - A copy of 29 CFR 1910.1043 and its Appendices;
  - A description of the affected employee's duties as they relate to the employee's exposure;
  - The employee's exposure level or anticipated exposure level;
  - A description of any personal protective equipment used or to be used; and
  - Information from previous medical examinations of the affected employee which is not readily available to the examining physician.

- ❑ The employer shall obtain and furnish the employee with a copy of a written opinion from the examining physician containing the following:
  - The results of the medical examination and tests including the FEV1, FVC, and FEV1/FVC ratio;
  - The physician's opinion as to whether the employee has any detected medical conditions which would place the employee at increased risk of material impairment of the employee's health from exposure to cotton dust;
  - The physician's recommended limitations upon the employee's exposure to cotton dust or upon the employee's use of respirators, including a determination of whether an employee can wear a negative-pressure respirator, and, where the employee cannot, a determination of the employee's ability to wear a powered air-purifying respirator; and
  - A statement that the employee has been informed by the physician of the results of the medical examination and any medical conditions which require further examination or treatment.

- ❑ The written opinion obtained by the employer shall not reveal specific findings or diagnoses unrelated to occupational exposure.

## Communication of Hazards to Employees

- ❑ The employer shall provide a training program for all employees exposed to cotton dust and shall assure that each employee is informed of the following:
  - The acute and long-term health hazards associated with exposure to cotton dust;
  - The names and descriptions of jobs and processes that could result in exposure to cotton dust at or above the PEL;
  - The measures necessary to protect the employee from exposures in excess of the PEL;
  - The purpose, proper use, and limitations of respirators;

- The purpose for, and a description of, the medical surveillance program and other information which will aid exposed employees in understanding the hazards of cotton dust exposure; and
- The contents of 29 CFR 1910.1043 and its appendices.

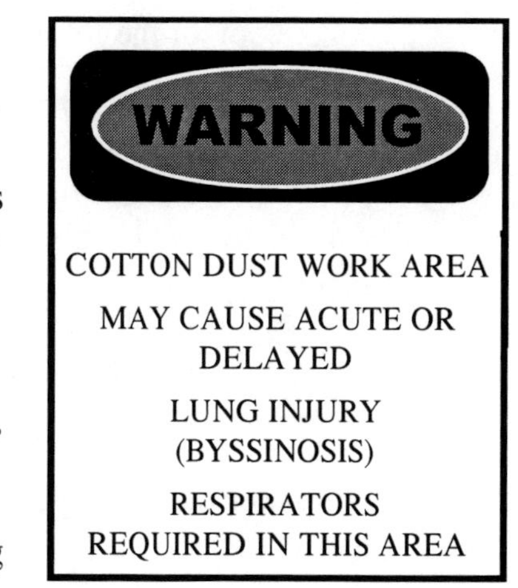


- The training program shall be provided prior to initial assignment and shall be repeated annually for each employee exposed to cotton dust, when job assignments or work processes change, and when employee performance indicates a need for retraining.
- Each employer shall post a copy of 29 CFR 1910.1043 with its appendices in a public location at the workplace, and shall, upon request, make copies available to employees.
- The employer shall provide upon request all materials relating to the employee training and information program to the Assistant Secretary and the Director.
- The employer shall post a warning sign, like the sample at right, in each work area where the PEL for cotton dust is exceeded.

## Recordkeeping

- The employer shall establish and maintain an accurate record of all measurements.
- The record shall include:
  - A log containing the items, dates, number, duration, and results of each of the samples taken, including a description of the procedure used to determine representative employee exposure;
  - The type of protective devices worn, if any, and length of time worn; and
  - The names, social security numbers, job classifications, and exposure levels of employees whose exposure the measurement is intended to represent.
- The employer shall maintain this record for at least 20 years.
- The employer shall establish and maintain an accurate medical record for each employee subject to medical surveillance.
- The record shall include:
  - The name, social security number, and description of the duties of the employee;
  - A copy of the medical examination results, including the medical history, questionnaire response, results of all tests, and the physician's recommendation;
  - A copy of the physician's written opinion;
  - Any employee medical complaints related to exposure to cotton dust;

- A copy of 29 CFR 1910.1043 and its appendices, except that the employer may keep one copy of the standard and its appendices for all employees, provided that he references the standard and appendices in the medical surveillance record of each employee; and
- A copy of the information provided to the physician.

- ❑ The employer shall maintain this record for at least 20 years.

- ❑ The employer shall make all records available to the Assistant Secretary and the Director for examination and copying.

- ❑ Employee exposure measurement records and employee medical records required by this paragraph shall be provided upon request to employees, designated representatives, and the Assistant Secretary in accordance with 29 CFR 1910.20(a) through (e) and (g) through (i).

- ❑ Whenever the employer ceases to do business, the successor employer shall receive and retain all records.

- ❑ Whenever the employer ceases to do business and there is no successor employer to receive and retain the records for the prescribed period, these records shall be transmitted to the Director.

- ❑ At the expiration of the retention period for the records required to be maintained by this section, the employer shall notify the Director at least three months prior to the disposal of such records and shall transmit those records to the Director if the Director requests them within that period.

- ❑ The employer shall also comply with any additional requirements involving transfer of records set forth in 29 CFR 1910.20(h).

## Observation of Monitoring

- ❑ The employer shall provide affected employees or their designated representatives an opportunity to observe any measuring or monitoring of employee exposure to cotton dust conducted.

- ❑ Whenever observation of the measuring or monitoring of employee exposure to cotton dust requires entry into an area where the use of personal protective equipment is required, the employer shall provide the observer with, and assure the use of, such equipment, and shall require the observer to comply with all other applicable safety and health procedures.

- ❑ Without interfering with the measurement, observers shall be entitled to:
  - An explanation of the measurement procedures;
  - An opportunity to observe all steps related to the measurement of airborne concentrations of cotton dust performed at the place of exposure; and
  - An opportunity to record the results obtained.

# 1,2-dibromo-3-chloropropane (DBCP)

## 1910.1044

Section 9 applies to occupational exposure to 1,2-dibromo-3-chloropropane (DBCP). This section does not apply to exposure that results solely from the application and use of DBCP as a pesticide or the storage, transportation, distribution, or sale of DBCP in intact containers sealed in such a manner as to prevent exposure to DBCP vapors or liquid.

## Definitions

- *Assistant Secretary* means the Assistant Secretary of Labor for Occupational Safety and Health, U.S. Department of Labor, or designee.
- *Authorized person* means any person required by his duties to be present in regulated areas and authorized to do so by his employer, this section, or the Act. Authorized person also includes any person entering such areas as a designated representative of employees exercising an opportunity to observe employee exposure monitoring.
- *DBCP* means 1,2-dibromo-3-chloropropane, CAS No. 96-12-8, and includes all forms of DBCP.
- *Director* means the Director, National Institute for Occupational Safety and Health, U.S. Department of Health and Human Services, or designee.
- *Emergency* means any occurrence such as, but not limited to, equipment failure, rupture of containers, or failure of control equipment that may, or does, result in an unexpected release of DBCP.
- *OSHA Area Office* means the Area Office of the Occupational Safety and Health Administration having jurisdiction over the geographic area where the affected workplace is located.

## Permissible Exposure Limit (PEL)

- ❑ The employer shall assure that no employee is exposed to an airborne concentration of DBCP in excess of 1 part DBCP per billion parts of air (ppb) as an 8-hour TWA.

- ❑ The employer shall assure that no employee is exposed to eye or skin contact with DBCP.

## Notification of Use

- ❑ Within ten days following the introduction of DBCP into the workplace, every employer who has a workplace where DBCP is present shall report the following information to the nearest OSHA Area Office for each such workplace:
  - The address and location of the workplace;
  - A brief description of each process or operation which may result in employee exposure to DBCP;
  - The number of employees engaged in each process or operation who may be exposed to DBCP, and an estimate of the frequency and degree of exposure that occurs; and
  - A brief description of the employer's safety and health program as it relates to the limitation of employee exposure to DBCP.

## Regulated Areas

- ❑ Within each place of employment, the employer shall establish regulated areas wherever DBCP concentrations are in excess of the PEL.

- ❑ The employer shall limit access to regulated areas to authorized persons.

## Exposure Monitoring

- ❑ Determinations of airborne exposure levels shall be made from air samples that are representative of each employee's exposure to DBCP over an 8-hour period.

- ❑ Employee exposure is that exposure which would occur if the employee were not using a respirator.

- ❑ Each employer who has a place of employment in which DBCP is present, shall monitor each workplace and work operation to accurately determine the airborne concentrations of DBCP to which employees may be exposed.

- ❑ If the monitoring required by this section reveals employee exposures to be below the PEL, the employer shall repeat these measurements at least quarterly.

- ❑ If the monitoring required by this section reveals employee exposures to be in excess of the PEL, the employer shall repeat these measurements for each such employee at least monthly.

The employer shall continue monthly monitoring until at least two consecutive measurements, taken at least seven days apart, are below the PEL. Thereafter the employer shall monitor at least quarterly.

- Whenever there has been a production, process, control, or personnel change which may result in any new or additional exposure to DBCP, or whenever the employer has any reason to suspect new or additional exposures to DBCP, the employer shall monitor the employees potentially affected by such change for the purpose of determining their current exposure.

- Within five working days after the receipt of the monitoring results, the employer shall notify each employee in writing of the measurements that represent the employee's exposure.

- Whenever the results indicate that employee exposure exceeds the PEL, the employer shall include in the written notice a statement that the PEL was exceeded and a description of the corrective action being taken to reduce exposure to or below the PEL.

- The employer shall use a method of measurement that has an accuracy, to a confidence level of 95 percent, of not less than plus or minus 25 percent for concentrations of DBCP at or above the PEL.

## Methods of Compliance

- The employer shall institute engineering and work-practice controls to reduce and maintain employee exposures to DBCP at or below the PEL, except to the extent that the employer establishes that such controls are not feasible. Where feasible engineering and work-practice controls are not sufficient to reduce employee exposures to within the PEL, the employer shall use them to reduce exposures to the lowest level achievable by these controls and shall supplement them by use of respiratory protection.

- The employer shall establish and implement a written program to reduce employee exposures to DBCP to or below the PEL solely by means of engineering and work-practice controls.

- The written program shall include a detailed schedule for development and implementation of the engineering and work-practice controls. These plans shall be revised at least every six months to reflect the current status of the program.

- Written plans for these compliance programs shall be submitted, upon request, to the Assistant Secretary and the Director, and shall be available at the worksite for examination and copying by the Assistant Secretary, the Director, and any affected employee or designated representative of employees.

- The employer shall institute and maintain at least the controls that are described in his most recent written compliance program.

## Respiratory Protection

- ☐ For employees who are required to use respirators by this section, the employer must provide respirators that comply with the requirements of this paragraph. Respirators must be used during:
  - Periods necessary to install or implement feasible engineering and work-practice controls.
  - Maintenance and repair activities for which engineering and work-practice controls are not feasible.
  - Work operations for which feasible engineering and work-practice controls are not yet sufficient to reduce employee exposure to or below the PEL.
  - Emergencies.
- ☐ The employer must implement a respiratory protection program in accordance with 29 CFR 1910.134 (b) through (d) (except (d)(1)(iii)) and (f) through (m).
- ☐ The employer must select the appropriate respirator from Table 4-9-1.

## Table 4-9-1: Respiratory Protection for DBCP

| Airborne concentration of DBCP or condition of use | Required Respirator |
|---|---|
| Less than or equal to 10 ppb | Any supplied-air respirator; or any self-contained breathing apparatus. |
| Less than or equal to 50 ppb | Any supplied-air respirator with full facepiece, helmet, or hood; or any self-contained breathing apparatus with full facepiece. |
| Less than or equal to 1,000 ppb | A Type C supplied-air respirator operated in pressure-demand or other positive-pressure or continuous flow mode. |
| Less than or equal to 2,000 ppb | A Type C supplied-air respirator with full facepiece operated in pressure-demand or other positive-pressure mode, or with full facepiece, helmet, or hood operated in continuous flow mode. |
| Greater than 2,000 ppb or inknown | A combination respirator which includes a Type C supplied-air respirator with full facepiece operated in pressure-demand or other positive-pressure or continuous flow mode, and an auxiliary self-contained breathing apparatus operated in pressure-demand or positive-pressure mode; or |
| | A self-contained breathing apparatus with full facepiece operated in pressure-demand or other positive-pressure mode. |
| Firefighting | A self-contained breathing apparatus with full facepiece operated in pressure-demand or other positive-pressure mode. |

## Emergency Situations

- A written plan for emergency situations shall be developed for each workplace in which DBCP is present.
- Appropriate portions of the plan shall be implemented in the event of an emergency.
- Employees engaged in correcting emergency conditions shall be properly equipped until the emergency is abated.
- Employees not engaged in correcting the emergency shall be removed and restricted from the area, and normal operations in the affected area shall not be resumed until the emergency is abated.
- Where there is a possibility of employee exposure to DBCP due to the occurrence of an emergency, a general alarm shall be installed and maintained to promptly alert employees of such occurrences.
- For any employee exposed to DBCP in an emergency situation, the employer shall provide medical surveillance.
- Following an emergency, the employer shall conduct monitoring.
- In workplaces not normally subject to periodic monitoring, the employer may terminate monitoring when two consecutive measurements indicate exposures below the PEL.

## Protective Clothing and Equipment

- Where there is any possibility of eye or dermal contact with liquid or solid DBCP, the employer shall provide, at no cost to the employee, and assure that the employee wears impermeable protective clothing and equipment to protect the area of the body which may come in contact with DBCP. Eye and face protection shall meet the requirements of 29 CFR 1910.133 of this part.
- The employer shall assure that employees only remove DBCP-contaminated work clothing in provided change rooms.
- The employer shall assure that employees promptly remove any protective clothing and equipment that becomes contaminated with DBCP-containing liquids and solids. This clothing shall not be reworn until the DBCP has been removed from the clothing or equipment.
- The employer shall assure that no employee takes DBCP-contaminated protective devices and work clothing out of the change room, except those employees authorized to do so for the purpose of laundering, maintenance, or disposal.
- DBCP-contaminated protective devices and work clothing shall be placed and stored in closed containers that prevent dispersion of the DBCP outside the container.

- ❑ Containers of DBCP-contaminated protective devices or work clothing, which are to be taken out of change rooms or the workplace for cleaning, maintenance, or disposal, shall bear labels.
- ❑ The employer shall clean, launder, repair, or replace protective clothing and equipment required by this paragraph to maintain their effectiveness. The employer shall provide clean protective clothing and equipment at least daily to each affected employee.
- ❑ The employer shall inform any person who launders or cleans DBCP-contaminated protective clothing or equipment of the potentially harmful effects of exposure to DBCP.
- ❑ The employer shall prohibit the removal of DBCP from protective clothing and equipment by blowing or shaking.

## Housekeeping

- ❑ All workplace surfaces shall be maintained free of visible accumulations of DBCP.
- ❑ Dry sweeping and the use of compressed air for the cleaning of floors and other surfaces is prohibited where DBCP dusts or liquids are present.
- ❑ Where vacuuming methods are selected to clean floors and other surfaces, either portable units or a permanent system may be used.
- ❑ If a portable unit is selected, the exhaust shall be attached to the general workplace exhaust ventilation system or collected within the vacuum unit and equipped with high-efficiency filters, or other appropriate means of contaminant removal, so that DBCP is not reintroduced into the workplace air.
- ❑ Portable vacuum units used to collect DBCP may not be used for other cleaning purposes and shall be labeled.
- ❑ Cleaning of floors and other surfaces contaminated with DBCP-containing dusts shall not be performed by washing with a hose, unless a fine spray has first been laid down.
- ❑ Where DBCP is present in a liquid form, or as a resultant vapor, all containers or vessels containing DBCP shall be enclosed to the maximum extent feasible and tightly covered when not in use.
- ❑ DBCP waste scrap, debris, containers, or equipment, shall be disposed of in sealed bags or other closed containers that prevent dispersion of DBCP outside the container.

## Hygiene Facilities and Practices

- The employer shall provide clean change rooms equipped with storage facilities for street clothes and separate storage facilities for protective clothing and equipment whenever employees are required to wear protective clothing and equipment.
- The employer shall assure employees working in regulated areas shower at the end of each work shift.
- The employer shall assure that employees whose skin becomes contaminated with DBCP-containing liquids or solids immediately wash or shower to remove any DBCP from the skin.
- The employer shall provide shower facilities in accordance with 29 CFR 1910.141(d)(3).
- The employer shall provide lunchroom facilities that have a temperature controlled, positive-pressure, filtered air supply, and which are accessible to employees working in regulated areas.
- The employer shall assure that employees working in regulated areas remove protective clothing and wash their hands and face prior to eating.
- The employer shall provide a sufficient number of lavatory facilities that comply with 29 CFR 1910.141(d)(1) and (2).
- The employer shall assure that, in regulated areas, food or beverages are not present or consumed, smoking products and implements are not present or used, and cosmetics are not present or applied.

## Medical Surveillance

- The employer shall make available a medical surveillance program for employees who work in regulated areas and employees who are subjected to DBCP exposures in an emergency situation.
- All medical examinations and procedures shall be performed by or under the supervision of a licensed physician, and shall be provided without cost to the employee.
- At the time of initial assignment, and annually thereafter, the employer shall provide a medical examination for employees who work in regulated areas, which includes at least the following:
  - A medical and occupational history, including reproductive history;
  - A physical examination, including examination of the genito-urinary tract, testicle size, and body habitus, including a determination of sperm count;
  - A serum specimen shall be obtained and the following determinations made by radioimmunoassay techniques utilizing National Institutes of Health (NIH) specific antigen or one of equivalent sensitivity:
    - Serum follicle stimulating hormone (FSH);

  - Serum luteinizing hormone (LH);
  - Serum total estrogen (females); and
  - Any other tests deemed appropriate by the examining physician.

- ❑ If the employee for any reason develops signs or symptoms commonly associated with exposure to DBCP, the employer shall provide the employee with a medical examination that shall include those elements considered appropriate by the examining physician.

- ❑ The employer shall provide the following information to the examining physician:
  - A copy of 29 CFR 1910.1044 and its appendices;
  - A description of the affected employee's duties as they relate to the employee's exposure;
  - The level of DBCP to which the employee is exposed; and
  - A description of any personal protective equipment used, or to be used.

- ❑ The employer shall obtain and provide the employee with a written opinion from the examining physician which shall include:
  - The results of the medical tests performed;
  - The physician's opinion as to whether the employee has any detected medical condition which would place the employee at an increased risk of material impairment of health from exposure to DBCP; and
  - Any recommended limitations upon the employee's exposure to DBCP or upon the use of protective clothing and equipment such as respirators.

- ❑ The employer shall instruct the physician not to reveal in the written opinion specific findings or diagnoses unrelated to occupational exposure.

- ❑ If the employee is exposed to DBCP in an emergency situation, the employer shall provide the employee with a sperm count test as soon as practicable, or, if the employee has been vasectionized or is unable to produce a semen specimen, the hormone tests. The employer shall provide these same tests three months later.

## Communication of Hazards to Employees

- ❑ The employer shall institute a training program for all employees who may be exposed to DBCP and shall assure their participation in such training program.

- ❑ The employer shall assure that each employee is informed of the following:
  - Information contained in 29 CFR 1910.1044 Appendix A;
  - The quantity, location, manner of use, release, and storage of DBCP, and the specific nature of operations which could result in exposure to DBCP as well as any necessary protective steps;
  - The purpose, proper use, and limitations of respirators; and
  - The purpose and description of the medical surveillance program.

- The employer shall provide, upon request, all materials relating to the employee information and training program to the Assistant Secretary and the Director.

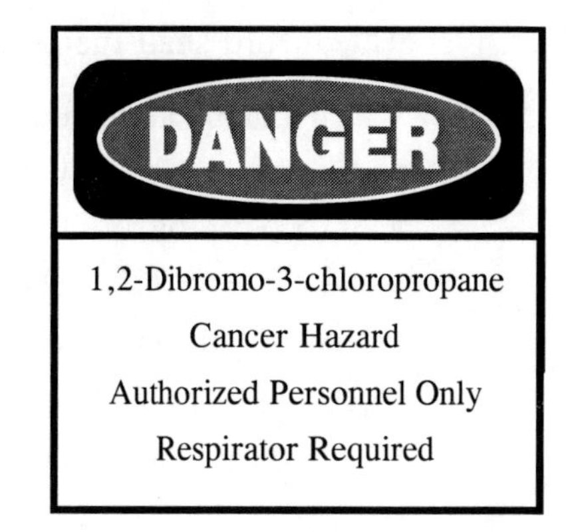


- The employer may use labels or signs required by other statutes, regulations, or ordinances in addition to, or in combination with, signs and labels.
- The employer shall assure that no statement appears on or near any sign or label that contradicts or detracts from the required sign or label.
- The employer shall post signs to clearly indicate all regulated areas. These signs shall look like the sample at above right.

- The employer shall assure that precautionary labels are affixed to all containers of DBCP and of products containing DBCP in the workplace, and that the labels remain affixed when the DBCP or products containing DBCP are sold, distributed, or otherwise leave the employer's workplace. Where DBCP or products containing DBCP are sold, distributed, or otherwise leave the employer's workplace bearing appropriate labels required by EPA under the regulations in 40 CFR Part 162, the labels required by this paragraph need not be affixed.
- The employer shall assure that the precautionary labels required by this paragraph are readily visible and legible. The labels shall look like the sample at below right.

## Recordkeeping

- The employer shall establish and maintain an accurate record of all monitoring.
- This record shall include:
  - The dates, number, duration, and results of each of the samples taken, including a description of the sampling procedure used to determine representative employee exposure;
  - A description of the sampling and analytical methods used;
  - Type of respiratory protective devices worn, if any; and
  - Name, social security number, and job classification of the employee monitored and of all other employees whose exposure the measurement is intended to represent.
- The employer shall maintain this record for at least 40 years or the duration of employment plus 20 years, whichever is longer.
- The employer shall establish and maintain an accurate record for each employee subject to medical surveillance.

- ❑ This record shall include:
  - The name and social security number of the employee;
  - A copy of the physician's written opinion;
  - Any employee medical complaints related to exposure to DBCP;
  - A copy of the information provided the physician; and
  - A copy of the employee's medical and work history.
- ❑ The employer shall maintain this record for at least 40 years or the duration of employment plus 20 years, whichever is longer.
- ❑ The employer shall assure that all records be made available, upon request, to the Assistant Secretary and the Director for examination and copying.
- ❑ Employee exposure monitoring records and employee medical records shall be provided, upon request to employees, designated representatives, and the Assistant Secretary in accordance with 29 CFR 1910.20 (a) through (e) and (g) through (i).
- ❑ If the employer ceases to do business, the successor employer shall receive and retain all records for the prescribed period.
- ❑ If the employer ceases to do business and there is no successor employer to receive and retain the records for the prescribed period, the employer shall transmit these records by mail to the Director.
- ❑ At the expiration of the retention period for the records, the employer shall transmit these records by mail to the Director.
- ❑ The employer shall also comply with any additional requirements involving transfer of records set forth in 29 CFR 1910.20(h).

## Observation of Monitoring

- ❑ The employer shall provide affected employees, or their designated representatives, with an opportunity to observe any monitoring of employee exposure to DBCP required by this section.
- ❑ Whenever observation of the measuring or monitoring of employee exposure to DBCP requires entry into an area where the use of protective clothing or equipment is required, the employer shall provide the observer with personal protective clothing or equipment required to be worn by employees working in the area, assure the use of such clothing and equipment, and require the observer to comply with all other applicable safety and health procedures.
- ❑ Without interfering with the monitoring or measurement, observers shall be entitled to:
  - Receive an explanation of the measurement procedures;
  - Observe all steps related to the measurement of airborne concentrations of DBCP performed at the place of exposure; and
  - Record the results obtained.

# Ethylene Oxide

## 1910.1047

Section 10 applies to all occupational exposures to ethylene oxide (EtO), CAS No. 75-21-8. This section does not apply to the processing, use, or handling of products containing EtO where objective data are reasonably relied upon that demonstrate the product is not capable of releasing EtO in airborne concentrations at or above the action level, and may not reasonably be foreseen to release EtO in excess of the excursion limit, under the expected conditions of processing, use, or handling that will cause the greatest possible release.

### Definitions

- *Action level* means a concentration of airborne EtO of 0.5 parts per million (ppm) calculated as an 8-hour time-weighted average (TWA).
- *Assistant Secretary* means the Assistant Secretary of Labor for Occupational Safety and Health, U.S. Department of Labor, or designee.
- *Authorized person* means any person specifically authorized by the employer whose duties require the person to enter a regulated area, or any person entering such an area as a designated representative of employees for the purpose of exercising the right to observe monitoring and measuring procedures, or any other person authorized by the Occupational Safety and Health Act.
- *Director* means the Director of the National Institute for Occupational Safety and Health, U.S. Department of Health and Human Services, or designee.
- *Emergency* means any occurrence such as, but not limited to, equipment failure, rupture of containers, or failure of control equipment that is likely to, or does, result in an unexpected significant release of EtO.

- *Employee exposure* means exposure to airborne EtO that would occur if the employee were not using respiratory protective equipment.

- *Ethylene oxide* or EtO means the three-membered ring organic compound with chemical formula $C_2H_4O$.

## Permissible Exposure Limits (PELs)

- The employer shall ensure that no employee is exposed to an airborne concentration of EtO in excess of 1 part EtO per million parts of air (1 ppm) as an 8-hour TWA.

- The employer shall ensure that no employee is exposed to an airborne concentration of EtO in excess of 5 parts of EtO per million parts of air (5 ppm) as averaged over a sampling period of 15 minutes.

## Exposure Monitoring

- Determinations of employee exposure shall be made from breathing zone air samples that are representative of the 8-hour TWA and 15-minute short-term exposures of each employee.

- Representative 8-hour TWA employee exposure shall be determined on the basis of one or more samples representing full-shift exposure for each shift for each job classification in each work area. Representative 15-minute short-term employee exposures shall be determined on the basis of one or more samples representing 15-minute exposures associated with operations that are most likely to produce exposures above the excursion limit for each shift for each job classification in each work area.

- Where the employer can document that exposure levels are equivalent for similar operations in different work shifts, the employer need only determine representative employee exposure for that operation during one shift.

- Each employer shall perform initial monitoring to determine accurately the airborne concentrations of EtO to which employees may be exposed.

- Where the employer has previously monitored for the excursion limit and the monitoring satisfies all other requirements of this section, the employer may rely on such earlier monitoring results to satisfy the requirements.

- If the monitoring reveals employee exposure at or above the action level but at or below the 8-hour TWA, the employer shall repeat such monitoring for each such employee at least every six months.

- If the monitoring reveals employee exposure above the 8-hour TWA, the employer shall repeat such monitoring for each such employee at least every three months.

- The employer may alter the monitoring schedule from quarterly to semiannually for any employee for whom two consecutive measurements taken at least seven days apart indicate that the employee's exposure has decreased to or below the 8-hour TWA.

- If the monitoring reveals employee exposure above the 15-minute excursion limit, the employer shall repeat such monitoring for each such employee at least every three months and more often, as necessary, to evaluate the employee's short-term exposures.

- If the initial monitoring reveals employee exposure to be below the action level, the employer may discontinue TWA monitoring for those employees whose exposures are represented by the initial monitoring.

- If the periodic monitoring reveals that employee exposures, as indicated by at least two consecutive measurements taken at least seven days apart, are below the action level, the employer may discontinue TWA monitoring for those employees whose exposures are represented by such monitoring.

- If the initial monitoring reveals employee exposure to be at or below the excursion limit, the employer may discontinue excursion limit monitoring for those employees whose exposures are represented by the initial monitoring.

- If the periodic monitoring reveals that employee exposures, as indicated by at least two consecutive measurements taken at least seven days apart, are at or below the excursion limit, the employer may discontinue excursion limit monitoring for those employees whose exposures are represented by such monitoring.

- The employer shall institute the exposure monitoring whenever there has been a change in the production, process, control equipment, personnel, or work practices that may result in new or additional exposures to EtO, or when the employer has any reason to suspect that a change may result in new or additional exposures.

- Monitoring shall be accurate, to a confidence level of 95 percent, to within plus or minus 25 percent for airborne concentrations of EtO at the 1 ppm TWA, and to within plus or minus 35 percent for airborne concentrations of EtO at the action level of 0.5 ppm.

- Monitoring shall be accurate, to a confidence level of 95 percent, to within plus or minus 35 percent for airborne concentrations of EtO at the excursion limit.

- The employer shall, within 15 working days after the receipt of the results of any monitoring performed under this standard, notify the affected employee of these results in writing either individually or by posting results in an appropriate location that is accessible to all affected employees.

- The written notification shall contain the corrective action being taken by the employer to reduce employee exposure to or below the TWA and/or excursion limit, wherever monitoring results indicated that the TWA and/or excursion limit has been exceeded.

## Regulated Areas

- The employer shall establish a regulated area wherever occupational exposure to airborne concentrations of EtO may exceed the TWA or the EtO concentration exceeds, or can reasonably be expected to exceed, the excursion limit.
- Access to regulated areas shall be limited to authorized persons.
- Regulated areas shall be demarcated in a manner that minimizes the number of employees within the regulated area.

## Methods of Compliance

- The employer shall institute engineering controls and work practices to reduce and maintain employee exposure to or below the TWA and excursion limit, except to the extent that such controls are not feasible.
- Wherever the feasible engineering controls and work practices that can be instituted are not sufficient to reduce employee exposure to or below the TWA and excursion limit, the employer shall use them to reduce employee exposure to the lowest levels achievable by these controls and shall supplement them by the use of respiratory protection.
- Engineering controls are generally infeasible for the following operations: collection of quality assurance sampling from sterilized materials; removal of biological indicators from sterilized materials; loading and unloading of tank cars; changing of ethylene oxide tanks on sterilizers; and vessel cleaning. For these operations, engineering controls are required only where the Assistant Secretary demonstrates that such controls are feasible.
- Where the TWA or excursion limit is exceeded, the employer shall establish and implement a written program to reduce exposure to or below the TWA and excursion limit by means of engineering and work-practice controls, and respiratory protection.
- The compliance program shall include a schedule for periodic leak detection surveys and a written plan for emergency situations.
- Written plans for a program shall be developed and furnished upon request for examination and copying to the Assistant Secretary, the Director, affected employees, and designated employee representatives. Such plans shall be reviewed at least every 12 months, and shall be updated as necessary to reflect significant changes in the status of the employer's compliance program.
- The employer shall not implement a schedule of employee rotation as a means of compliance with the TWA or excursion limit.

## Respiratory Protection

- For employees who use respirators, the employer must provide respirators that comply with the requirements of this paragraph. Respirators must be used during:
  - Periods necessary to install or implement feasible engineering and work-practice controls.
  - Work operations, such as maintenance and repair activities and vessel cleaning, for which engineering and work-practice controls are not feasible.
  - Work operations for which feasible engineering and work-practice controls are not yet sufficient to reduce employee exposure to or below the TWA.
  - Emergencies.
- The employer must implement a respiratory protection program in accordance with 29 CFR 1910.134(b) through (d) (except (d)(1)(iii)) and (f) through (m).
- The employer must select the appropriate respirator from Table 4-10-1.

## Protective Clothing and Equipment

- When employees could have eye or skin contact with EtO or EtO solutions, the employer must select and provide, at no cost to the employee, appropriate protective clothing or other equipment in accordance with 29 CFR 1910.132-133 to protect any area of the employee's body that may come in contact with the EtO or EtO solution, and must ensure that the employee wears the protective clothing and equipment provided.

### Table 4-10-1: Respiratory Protection for Airborne EtO

| Condition of use or concentration of airborne EtO (ppm) | Required respirator |
|---|---|
| Equal to or less than 50 | Full facepiece respirator with EtO-approved canister, front- or back-mounted. |
| Equal to or less than 2,000 | Positive-pressure supplied-air respirator equipped with full facepiece, hood, or helmet; or |
| | Continuous-flow supplied-air respirator (positive-pressure) equipped with hood, helmet, or suit. |
| Concentration above 2,000 unknown concentration, or emergencies | Positive-pressure self-contained breathing apparatus (SCBA), equipped with full facepiece; or |
| | Positive-pressure full facepiece supplied0-air respirator equipped with an auxiliary positive-pressure self-contained breathing apparatus. |
| Firefighting | Positive-pressure self-contained breathing apparatus equipped with full facepiece. |
| Escape | Any respirator described above. |

Note: Respirators approved for use in higher concentrations are permitted to be used in lower concentrations.

## Emergency Situations

- ❑ A written plan for emergency situations shall be developed for each workplace where there is a possibility of an emergency. Appropriate portions of the plan shall be implemented in the event of an emergency.

- ❑ The plan shall specifically provide that employees engaged in correcting emergency conditions be equipped with respiratory protection until the emergency is abated.

- ❑ The plan shall include the elements prescribed in 29 CFR 1910.38, "Employee emergency plans and fire prevention plans."

- ❑ Where there is the possibility of employee exposure to EtO due to an emergency, means shall be developed to alert potentially affected employees of such occurrences promptly. Affected employees shall be immediately evacuated from the area in the event that an emergency occurs.

## Medical Surveillance

- ❑ The employer shall institute a medical surveillance program for all employees who are or may be exposed to EtO at or above the action level, without regard to the use of respirators, for at least 30 days a year.

- ❑ The employer shall make available medical examinations and consultations to all employees who have been exposed to EtO in an emergency situation.

- ❑ The employer shall ensure that all medical examinations and procedures are performed by or under the supervision of a licensed physician, and are provided without cost to the employee, without loss of pay, and at a reasonable time and place.

- ❑ The employer shall make available medical examinations and consultations to each employee on the following schedules:
  - Prior to assignment of the employee to an area where exposure may be at or above the action level for at least 30 days a year.
  - At least annually for each employee exposed at or above the action level for at least 30 days in the past year.
  - At termination of employment or reassignment to an area where exposure to EtO is not at or above the action level for at least 30 days a year.
  - As medically appropriate for any employee exposed during an emergency.
  - As soon as possible upon notification by an employee either that the employee has developed signs or symptoms indicating possible overexposure to EtO, or the employee desires medical advice concerning the effects of current or past exposure to EtO on the employee's ability to produce a healthy child.
- ❑ If the examining physician determines that any of the examinations should be provided more frequently than specified, the employer shall provide such examinations to affected employees at the frequencies recommended by the physician.

- Medical examinations shall include:
  - A medical and work history with special emphasis directed to symptoms related to the pulmonary, hematologic, neurologic, and reproductive systems and to the eyes and skin.
  - A physical examination with particular emphasis given to the pulmonary, hematologic, neurologic, and reproductive systems and to the eyes and skin.
  - A complete blood count to include at least a white cell count (including differential cell count), red cell count, hematocrit, and hemoglobin.
  - Any laboratory or other test which the examining physician deems necessary by sound medical practice.
- The content of medical examinations or consultation made available shall be determined by the examining physician, and shall include pregnancy testing or laboratory evaluation of fertility, if requested by the employee and deemed appropriate by the physician.
- The employer shall provide the following information to the examining physician:
  - A copy of 29 CFR 1910.1047 and Appendices A, B, and C;
  - A description of the affected employee's duties as they relate to the employee's exposure;
  - The employee's representative exposure level or anticipated exposure level;
  - A description of any personal protective and respiratory equipment used, or to be used; and
  - Information from previous medical examinations of the affected employee that is not otherwise available to the examining physician.
- The employer shall obtain a written opinion from the examining physician. This written opinion shall contain the results of the medical examination and shall include:
  - The physician's opinion as to whether the employee has any detected medical conditions that would place the employee at an increased risk of material health impairment from exposure to EtO;
  - Any recommended limitations on the employee or upon the use of personal protective equipment such as clothing or respirators; and
  - A statement that the employee has been informed by the physician of the results of the medical examination and of any medical conditions resulting from EtO exposure that require further explanation or treatment.
- The employer shall instruct the physician not to reveal in the written opinion given to the employer specific findings or diagnoses unrelated to occupational exposure to EtO.
- The employer shall provide a copy of the physician's written opinion to the affected employee within 15 days from its receipt.

## Communication of Hazards to Employees

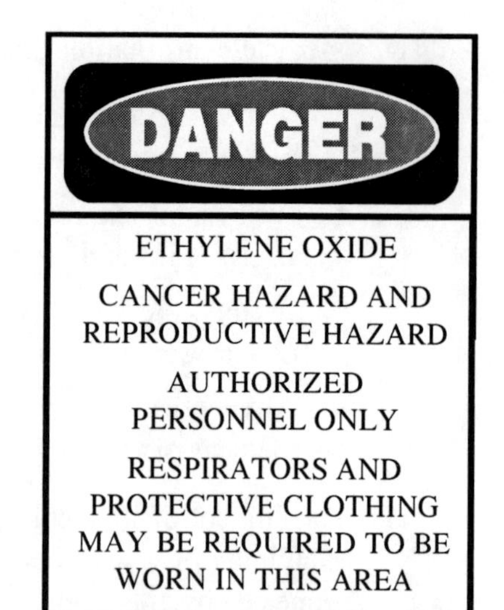


- ❑ The employer shall post and maintain legible signs demarcating regulated areas and entrances or accessways to regulated areas These signs shall look like the sample at above right.

- ❑ The employer shall ensure that precautionary labels are affixed to all containers of EtO whose contents are capable of causing employee exposure at or above the action level or whose contents may reasonably be foreseen to cause employee exposure above the excursion limit. The employer shall ensure that the labels remain affixed when the containers of EtO leave the workplace. Reaction vessels, storage tanks, and pipes or piping systems are not considered to be containers. The labels shall comply with the requirements of 29 CFR 1910.1200(f) of OSHA's Hazard Communication Standard, and shall look like the sample at below right.

- ❑ The labeling requirements do not apply where EtO is used as a pesticide, as such term is defined in the Federal Insecticide, Fungicide, and Rodenticide Act (7 U.S.C. 136 et seq.), and it is labeled pursuant to that Act and regulations issued under that Act by the Environmental Protection Agency.

- ❑ Employers who are manufacturers or importers of EtO shall comply with the requirements regarding development of material safety data sheets as specified in 29 CFR 1910.1200(g) of OSHA's Hazard Communication Standard.

- ❑ The employer shall provide employees who are potentially exposed to EtO at or above the action level or above the excursion limit with information and training on EtO at the time of initial assignment and at least annually thereafter.

- ❑ Employees shall be informed of the following:
  - The requirements of 29 CFR 1910.1047 with an explanation of its contents, including Appendices A and B;
  - Any operations in their work area where EtO is present;
  - The location and availability of the written EtO final rule; and
  - The medical surveillance program with an explanation of the information in 29 CFR 1910.1047, Appendix C.

- ❑ Employee training shall include at least:
  - Methods and observations that may be used to detect the presence or release of EtO in the work area (such as monitoring conducted by the employer, continuous monitoring devices, etc.);

- The physical and health hazards of EtO;
- The measures employees can take to protect themselves from hazards associated with EtO exposure, including specific procedures the employer has implemented to protect employees from exposure to EtO, such as work practices, emergency procedures, and personal protective equipment to be used; and
- The details of the hazard communication program developed by the employer, including an explanation of the labeling system and how employees can obtain and use the appropriate hazard information.

## Recordkeeping

- Records shall include at least the following information:
  - The product qualifying for exemption;
  - The source of the objective data;
  - The testing protocol, results of testing, and/or analysis of the material for the release of EtO;
  - A description of the operation exempted and how the data support the exemption; and
  - Other data relevant to the operations, materials, processing, or employee exposures covered by the exemption.
- The employer shall maintain this record for the duration of the employer's reliance upon such objective data.
- The employer shall keep an accurate record of all measurements taken to monitor employee exposure to EtO.
- This record shall include at least the following information:
  - The date of measurement;
  - The operation involving exposure to EtO which is being monitored;
  - Sampling and analytical methods used and evidence of their accuracy;
  - Number, duration, and results of samples taken;
  - Type of protective devices worn, if any; and
  - Name, social security number, and exposure of the employees whose exposures are represented.
- The employer shall maintain this record for at least 30 years, in accordance with 29 CFR 1910.20.
- The employer shall establish and maintain an accurate record for each employee subject to medical surveillance, in accordance with 29 CFR 1910.20.

- ❑ The record shall include at least the following information:
  - The name and social security number of the employee;
  - The physicians' written opinions;
  - Any employee medical complaints related to exposure to EtO; and
  - A copy of the information provided to the physician.
- ❑ The employer shall ensure that this record is maintained for the duration of employment plus 30 years, in accordance with 29 CFR 1910.20.
- ❑ The employer shall make all records required to be maintained by this section available, upon request, to the Assistant Secretary and the Director for examination and copying.
- ❑ The employer shall make any exemption and exposure records available, upon request, for examination and copying to affected employees, former employees, designated representatives, and the Assistant Secretary, in accordance with 29 CFR 1910.20 (a) through (e) and (g) through (i).
- ❑ The employer shall make employee medical records available, upon request, for examination and copying to the subject employee, anyone having the specific written consent of the subject employee, and the Assistant Secretary, in accordance with 29 CFR 1910.20.
- ❑ The employer shall comply with the requirements concerning transfer of records set forth in 29 CFR 1910.20(h).
- ❑ Whenever the employer ceases to do business and there is no successor employer to receive and retain the records for the prescribed period, the employer shall notify the Director at least 90 days prior to disposal and transmit these records to the Director.

## Observation of Monitoring

- ❑ The employer shall provide affected employees or their designated representatives an opportunity to observe any monitoring conducted of employee exposure to EtO.
- ❑ When observation of the monitoring of employee exposure to EtO requires entry into an area where the use of protective clothing or equipment is required, the observer shall be provided with, and be required to use, such clothing and equipment and shall comply with all other applicable safety and health procedures.

# Formaldehyde

## 1910.1048

Section 11 applies to all occupational exposures to formaldehyde (i.e. from formaldehyde gas, its solutions, and materials that release formaldehyde).

## Definitions

- *Action level* means a concentration of 0.5 part formaldehyde per million parts of air (0.5 ppm) calculated as an 8-hour time-weighted average (TWA) concentration.
- *Assistant Secretary* means the Assistant Secretary of Labor for the Occupational Safety and Health Administration, U.S. Department of Labor, or designee.
- *Authorized person* means any person required by work duties to be present in regulated areas, or authorized to do so by the employer, this section, or the OSH Act of 1970.
- *Director* means the Director of the National Institute for Occupational Safety and Health, U.S. Department of Health and Human Services, or designee.
- *Emergency* is any occurrence, such as, but not limited to, equipment failure, rupture of containers, or failure of control equipment that results in an uncontrolled release of a significant amount of formaldehyde.
- *Employee exposure* means the exposure to airborne formaldehyde which would occur without the use of respiratory protection.
- *Formaldehyde* means the chemical substance, HCHO, CAS No. 50-00-0.

## Permissible Exposure Limit (PEL)

- ❑ The employer shall assure that no employee is exposed to an airborne concentration of formaldehyde that exceeds 0.75 parts formaldehyde per million parts of air (0.75 ppm) as an 8-hour TWA.

- ❑ The employer shall assure that no employee is exposed to an airborne concentration of formaldehyde that exceeds 2 parts formaldehyde per million parts of air (2 ppm) as a 15-minute STEL.

## Exposure Monitoring

- ❑ Each employer shall monitor employees to determine their exposure to formaldehyde.

- ❑ Where the employer documents, using objective data, that the presence of formaldehyde or formaldehyde-releasing products in the workplace cannot result in airborne concentrations of formaldehyde that would cause any employee to be exposed at or above the action level or the STEL under foreseeable conditions of use, the employer will not be required to measure employee exposure to formaldehyde.

- ❑ When an employee's exposure is determined from representative sampling, the measurements used shall be representative of the employee's full shift or short-term exposure to formaldehyde, as appropriate.

- ❑ Representative samples for each job classification in each work area shall be taken for each shift unless the employer can document with objective data that exposure levels for a given job classification are equivalent for different work shifts.

- ❑ The employer shall identify all employees who may be exposed at or above the action level or at or above the STEL and accurately determine the exposure of each employee so identified.

- ❑ Unless the employer chooses to measure the exposure of each employee potentially exposed to formaldehyde, the employer shall develop a representative sampling strategy and measure sufficient exposures within each job classification for each work shift to correctly characterize and not underestimate the exposure of any employee within each exposure group.

- ❑ The initial monitoring process shall be repeated each time there is a change in production, equipment, process, personnel, or control measures which may result in new or additional exposure to formaldehyde.

- ❑ If the employer receives reports of signs or symptoms of respiratory or dermal conditions associated with formaldehyde exposure, the employer shall promptly monitor the affected employee's exposure.

- ❑ The employer shall periodically measure and accurately determine exposure to formaldehyde for employees shown by the initial monitoring to be exposed at or above the action level or at or above the STEL.

- If the last monitoring results reveal employee exposure at or above the action level, the employer shall repeat monitoring of the employees at least every six months.

- If the last monitoring results reveal employee exposure at or above the STEL, the employer shall repeat monitoring of the employees at least once a year under worst conditions.

- The employer may discontinue periodic monitoring for employees if results from two consecutive sampling periods taken at least seven days apart show that employee exposure is below the action level and the STEL. The results must be statistically representative and consistent with the employer's knowledge of the job and work operation.

- Monitoring shall be accurate, at the 95 percent confidence level, to within plus or minus 25 percent for airborne concentrations of formaldehyde at the TWA and the STEL and to within plus or minus 35 percent for airborne concentrations of formaldehyde at the action level.

- The following guidelines shall be used for employee notification of monitoring results:
  - Within 15 days of receiving the results of exposure monitoring conducted under this standard, the employer shall notify the affected employees of these results.
  - Notification shall be in writing, either by distributing copies of the results to the employees or by posting the results.
  - If the employee exposure is over either PEL, the employer shall develop and implement a written plan to reduce employee exposure to or below both PELs, and give written notice to employees.
  - The written notice shall contain a description of the corrective action being taken by the employer to decrease exposure.

- The employer shall provide affected employees or their designated representatives an opportunity to observe any monitoring of employee exposure to formaldehyde required by this standard.

- When observation of the monitoring of employee exposure to formaldehyde requires entry into an area where the use of protective clothing or equipment is required, the employer shall provide the clothing and equipment to the observer, require the observer to use such clothing and equipment, and assure that the observer complies with all other applicable safety and health procedures.

## Regulated Areas

- The employer shall establish regulated areas where the concentration of airborne formaldehyde exceeds either the TWA or the STEL and post all entrances and access ways with signs similar to the sample at right.

- The employer shall limit access to regulated areas to authorized persons who have been trained to recognize the hazards of formaldehyde.

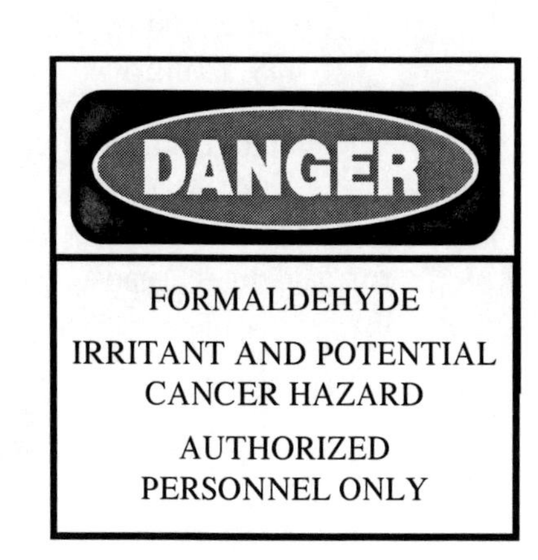

- ❑ An employer at a multi-employer worksite who establishes a regulated area shall communicate the access restrictions and locations of these areas to other employers with work operations at that worksite.

## Methods of Compliance

- ❑ The employer shall institute engineering and work-practice controls to reduce and maintain employee exposures to formaldehyde at or below the TWA and the STEL.

- ❑ Whenever the employer has established that feasible engineering and work-practice controls cannot reduce employee exposure to or below either of the PELs, the employer shall apply controls to reduce employee exposures to the extent feasible and shall supplement them with respirators which satisfy this standard.

## Respiratory Protection

- ❑ Respirators must be used during:
  - Periods necessary to install or implement feasible engineering and work-practice controls.
  - Work operations, such as maintenance and repair activities or vessel cleaning, for which the employer establishes that engineering and work-practice controls are not feasible.
  - Work operations for which feasible engineering and work-practice controls are not yet sufficient to reduce employee exposure to or below the PELs.
  - Emergencies.

- ❑ The employer must implement a respiratory protection program in accordance with 29 CFR 1910.134(b) through (d) (except (d)(1)(iii), and (d)(3)(iii)(B)(1) and (2)) and (f) through (m).

- ❑ If air-purifying chemical-cartridge respirators are used, the employer must:
  - Replace the cartridge after three hours of use or at the end of the workshift, whichever occurs first, unless the cartridge contains a NIOSH-approved end-of-service-life indicator (ESLI) to show when breakthrough occurs.
  - Replace canisters used in atmospheres up to 7.5 ppm (10 X PEL) every four hours, and industrial-sized canisters used in atmospheres up to 75 ppm (100 X PEL) every two hours, or at the end of the workshift, whichever occurs first, unless the canister contains a NIOSH-approved ESLI to show when breakthrough occurs.

- ❑ The employer must select appropriate respirators from Table 4-11-1.

- ❑ The employer must provide a powered air-purifying respirator adequate to protect against formaldehyde exposure to any employee who has difficulty using a negative-pressure respirator.

## Table 4-11-1: Respiratory Protection for Formaldehyde

| Condition of use or formaldehyde concentration (ppm) | Minimum respirator required[1] |
|---|---|
| Up to 7.5 ppm (10 x PEL) | Full facepiece with cartridges or canisters specifically approved for protection against formaldehyde.[2] |
| Up to 75 ppm (100 x PEL) | Full-face mask with chin style, or chest- or back-mounted type, with industrial-size canister specifically approved for protection against formaldehyde, or Type C supplied-air respirator, demand type, or continuous flow type, with full facepiece, hood, or helmet. |
| Above 75 ppm or unknown (emergencies) (100 x PEL) | Self-contained breathing apparatus (SCBA) with positive-pressure full facepiece, or Combination supplied-air, full facepiece, positive-pressure respirator with auxiliary self-contained air supply. |
| Firefighting | SCBA with positive-pressure full facepiece. |
| Escape | SCBA in demand or pressure-demand mode. Full-face mask with chin style, or front- or back-mounted type industrial-size canister specifically approved for protection against formaldehyde. |

[1] Respirators specified for use at higher concentrations may be used at lower concentrations.
[2] A half-mask respirator with cartridges specifically approved for protection against formaldehyde can be substituted for the full-facepiece respirator providing that effective gas-proof goggles are provided and used in combination with the half-mask respirator.

## Protective Clothing and Equipment

- Employers shall comply with the provisions of 29 CFR 1910.132–133. When protective equipment or clothing is provided under these provisions, the employer shall provide these protective devices at no cost to the employee, and assure that the employee wears them.
- The employer shall select protective clothing and equipment based upon the form of formaldehyde to be encountered, the conditions of use, and the hazard to be prevented.
- All contact of the eyes and skin with liquids containing 1 percent or more formaldehyde shall be prevented by the use of chemical protective clothing made of material impervious to formaldehyde and the use of other personal protective equipment, such as goggles and face shields, as appropriate to the operation.
- Contact with irritating or sensitizing materials shall be prevented to the extent necessary to eliminate the hazard.
- Where a face shield is worn, chemical safety goggles are also required if there is a danger of formaldehyde reaching the area of the eye.

- ❑ Full-body protection shall be worn for entry into areas where concentrations exceed 100 ppm and for emergency reentry into areas of unknown concentration.

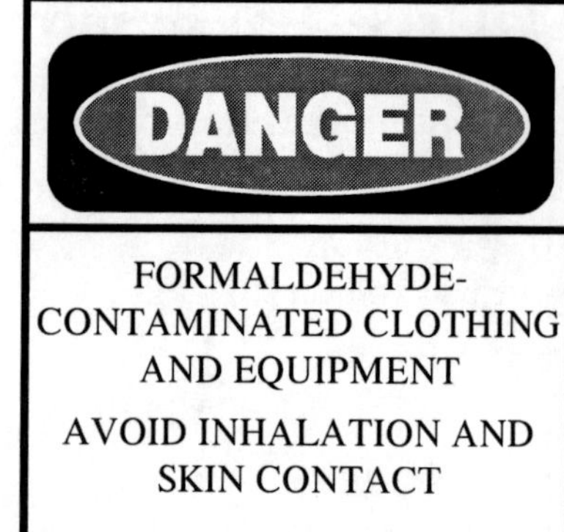


- ❑ The employer shall assure that protective equipment and clothing that has become contaminated with formaldehyde is cleaned or laundered before its reuse.

- ❑ When ventilating formaldehyde-contaminated clothing and equipment, the employer shall establish a storage area so that employee exposure is minimized. Containers for contaminated clothing and equipment and storage areas shall have labels and signs similar to the sample at right.

- ❑ The employer shall assure that only persons trained to recognize the hazards of formaldehyde remove the contaminated material from the storage area for purposes of cleaning, laundering, or disposal.

- ❑ The employer shall assure that no employee takes home equipment or clothing that is contaminated with formaldehyde.

- ❑ The employer shall repair or replace all required protective clothing and equipment for each affected employee as necessary to assure its effectiveness.

- ❑ The employer shall inform any person who launders, cleans, or repairs such clothing or equipment of formaldehyde's potentially harmful effects and of procedures to safely handle the clothing and equipment.

## Hygiene Facilities and Practices

- ❑ The employer shall provide change rooms, as described in 29 CFR 1910.141, for employees who are required to change from work clothing into protective clothing to prevent skin contact with formaldehyde.

- ❑ Because an employee's skin may be splashed with solutions containing 1 percent or greater formaldehyde, for example, because of equipment failure or improper work practices, the employer shall provide conveniently located quick drench showers and assure that affected employees use these facilities immediately.

- ❑ If there is any possibility that an employee's eyes may be splashed with solutions containing 0.1 percent or greater formaldehyde, the employer shall provide acceptable eyewash facilities within the immediate work area for emergency use.

## Housekeeping

- For operations involving formaldehyde liquids or gas, the employer shall conduct a program to detect leaks and spills, including regular visual inspections.
- Preventative maintenance of equipment, including surveys for leaks, shall be undertaken at regular intervals.
- In work areas where spillage may occur, the employer shall make provisions to contain the spill, decontaminate the work area, and dispose of the waste.
- The employer shall assure that all leaks are repaired and spills are cleaned promptly by employees wearing suitable protective equipment and trained in proper methods for cleanup and decontamination.
- Formaldehyde-contaminated waste and debris resulting from leaks or spills shall be placed for disposal in sealed containers bearing a label warning of formaldehyde's presence and of the hazards associated with formaldehyde.

## Emergency Situations

- For each workplace where there is the possibility of an emergency involving formaldehyde, the employer shall assure appropriate procedures are adopted to minimize injury and loss of life. Appropriate procedures shall be implemented in the event of an emergency.

## Medical Surveillance

- The employer shall institute medical surveillance programs for all employees exposed to formaldehyde at concentrations at or exceeding the action level or STEL.
- The employer shall make medical surveillance available for employees who develop signs and symptoms of overexposure to formaldehyde and for all employees exposed to formaldehyde in emergencies. When determining whether an employee may be experiencing signs and symptoms of possible overexposure to formaldehyde, the employer may rely on the evidence that signs and symptoms associated with formaldehyde exposure will occur only in exceptional circumstances when airborne exposure is less than 0.1 ppm and when formaldehyde is present in material in concentrations less than 0.1 percent.
- All medical procedures, including administration of medical disease questionnaires, shall be performed by or under the supervision of a licensed physician and shall be provided without cost to the employee, without loss of pay, and at a reasonable time and place.
- The employer shall make the following medical surveillance available to employees prior to assignment to a job where formaldehyde exposure is at or above the action level or above the STEL and annually thereafter. The employer shall also make the following medical surveillance available promptly upon determining that an employee is experiencing signs and symptoms indicative of possible overexposure to formaldehyde:

- Administer a medical disease questionnaire, which is designed to elicit information on work history; smoking history; any evidence of eye, nose, or throat irritation; chronic airway problems or hyperactive airway disease; allergic skin conditions or dermatitis; and upper or lower respiratory problems.

- A determination by the physician, based on evaluation of the medical disease questionnaire, of whether a medical examination is necessary for employees not required to wear respirators to reduce exposure to formaldehyde.

❑ Medical examinations shall be given to any employee who the physician feels, based on information in the medical disease questionnaire, may be at increased risk from exposure to formaldehyde, and, at the time of initial assignment and at least annually thereafter, to all employees required to wear a respirator to reduce exposure to formaldehyde. The medical examination shall include:

- A physical examination with emphasis on evidence of irritation or sensitization of the skin and respiratory system, shortness of breath, or irritation of the eyes.
- Laboratory examinations for respirator wearers consisting of baseline and annual pulmonary function tests. As a minimum, these tests shall consist of forced vital capacity (FVC), forced expiratory volume in one second (FEV1), and forced expiratory flow (FEF).
- Any other test which the examining physician deems necessary to complete the written opinion.
- Counseling of employees having medical conditions that would be directly or indirectly aggravated by exposure to formaldehyde or the increased risk of impairment of their health.

❑ The employer shall make medical examinations available as soon as possible to all employees who have been exposed to formaldehyde in an emergency. The medical examination shall include:

- A medical and work history with emphasis on any evidence of upper or lower respiratory problems, allergic conditions, skin reaction or hypersensitivity, and any evidence of eye, nose, or throat irritation; and
- Other elements considered appropriate by the examining physician.

❑ The employer shall provide the following information to the examining physician:

- A copy of 29 CFR 1910.1048 and Appendices A, C, D, and E;
- A description of the affected employee's job duties as they relate to the employee's exposure to formaldehyde;
- The representative exposure level for the employee's job assignment;
- Information concerning any personal protective equipment and respiratory protection used or to be used, by the employee.
- Information from previous medical examinations of the affected employee within the control of the employer; and
- In the event of a non-routine examination because of an emergency, a description of how the emergency occurred and the exposure the victim may have received.

- The employer shall obtain a written opinion from the examining physician. This written opinion shall contain the results of the medical examination except findings or diagnoses unrelated to occupational exposure to formaldehyde. The written opinion shall include:
  - The physician's opinion as to whether the employee has any medical condition that would place the employee at an increased risk of material impairment of health from exposure to formaldehyde;
  - Any recommended limitations on the employee's exposure or changes in the use of personal protective equipment, including respirators; and
  - A statement that the employee has been informed by the physician of any medical conditions which would be aggravated by exposure to formaldehyde, whether these conditions may have resulted from past formaldehyde exposure or from exposure in an emergency, and whether there is a need for further examination or treatment.
- The employer shall ensure retention of the results of the medical examination and tests conducted by the physician.
- The employer shall provide a copy of the physician's written opinion to the affected employee within 15 days of its receipt.
- When an employee reports significant irritation of the mucosa of the eyes or the upper airways, respiratory sensitization, dermal irritation, or dermal sensitization attributed to workplace formaldehyde exposure, medical removal provisions do not apply in the case of dermal irritation or dermal sensitization when the product suspected of causing the dermal condition contains less than 0.05 percent formaldehyde.
- An employee's report of signs or symptoms of possible overexposure to formaldehyde shall be evaluated by a physician selected by the employer. If the physician determines that a medical examination is not necessary, there shall be a two-week evaluation and remediation period to permit the employer to ascertain whether the signs or symptoms subside untreated or with the use of creams, gloves, first-aid treatment, or personal protective equipment. Industrial hygiene measures that limit the employee's exposure to formaldehyde may also be implemented during this period. The employee shall be referred immediately to a physician prior to expiration of the two-week period if the signs or symptoms worsen. Earnings, seniority, and benefits may not be altered during the two-week period by virtue of the report.
- If the signs or symptoms have not subsided or been remedied by the end of the two-week period, or earlier if signs or symptoms warrant, the employee shall be examined by a physician selected by the employer. The physician shall presume, absent contrary evidence, that observed dermal irritation or dermal sensitization are not attributable to formaldehyde when products to which the affected employee is exposed contain less than 0.1 percent formaldehyde.
- If the physician finds that significant irritation of the mucosa of the eyes or the upper airways, respiratory sensitization, dermal irritation, or dermal sensitization result from workplace formaldehyde exposure and recommends restrictions or removal, the employer shall promptly comply with the restrictions or recommendation of removal. In the event of a recommendation of removal, the employer shall remove the affected employee from the current formaldehyde

exposure and, if possible, transfer the employee to work having no or significantly less exposure to formaldehyde.

- ❑ When an employee is removed, the employer shall transfer the employee to comparable work for which the employee is qualified, or can be trained in a short period (up to six months), where the formaldehyde exposures are as low as possible, but not higher than the action level. The employer shall maintain the employee's current earnings, seniority, and other benefits. If there is no such work available, the employer shall maintain the employee's current earnings, seniority, and other benefits until such work becomes available, the employee is determined to be unable to return to workplace formaldehyde exposure, the employee is determined to be able to return to the original job status, or for six months, whichever comes first.

- ❑ The employer shall arrange for a follow-up medical examination to take place within six months after the employee is removed pursuant to this paragraph. This examination shall determine if the employee can return to the original job status, or if the removal is to be permanent. The physician shall make a decision within six months of the date the employee was removed as to whether the employee can be returned to the original job status, or if the removal is to be permanent.

- ❑ An employer's obligation to provide earnings, seniority, and other benefits to a removed employee may be reduced to the extent that the employee receives compensation for earnings lost during the period of removal either from a publicly- or employer-funded compensation program or from employment with another employer made possible by virtue of the employee's removal.

- ❑ In making determinations of the formaldehyde content of materials, the employer may rely on objective data.

- ❑ After the employer selects the initial physician who conducts any medical examination or consultation to determine whether medical removal or restriction is appropriate, the employee may designate a second physician to review any findings, determinations, or recommendations of the initial physician and to conduct such examinations, consultations, and laboratory tests as the second physician deems necessary and appropriate to evaluate the effects of formaldehyde exposure and to facilitate this review.

- ❑ The employer shall promptly notify an employee of the right to seek a second medical opinion after each occasion that an initial physician conducts a medical examination or consultation for the purpose of medical removal or restriction.

- ❑ The employer may condition its participation in, and payment for, the multiple physician review mechanism upon the employee doing the following within 15 days of receipt of the notification of the right to seek a second medical opinion, or receipt of the initial physician's written opinion, whichever is later:
  - The employee informs the employer of the intention to seek a second medical opinion.
  - The employee initiates steps to make an appointment with a second physician.

- If the findings, determinations, or recommendations of the second physician differ from those of the initial physician, then the employer and the employee shall assure that efforts are made for the two physicians to resolve the disagreement. If the two physicians are unable to quickly resolve their disagreement, then the employer and the employee through their respective physicians shall designate a third physician, who shall be a specialist in the field at issue, who will:
  - Review the findings, determinations, or recommendations of the prior physicians; and
  - Conduct such examinations, consultations, laboratory tests, and discussions with the prior physicians as the third physician deems necessary to resolve the disagreement of the prior physicians.
- Alternatively, the employer and the employee, or authorized employee representative, may jointly designate such third physician.
- The employer shall act consistent with the findings, determinations, and recommendations of the third physician, unless the employer and the employee reach an agreement that is otherwise consistent with the recommendations of at least one of the three physicians.

## Communication of Hazards to Employees

- The following shall be subject to the hazard communication requirements: formaldehyde gas, all mixtures or solutions composed of greater than 0.1 percent formaldehyde, and materials capable of releasing formaldehyde into the air, under reasonably foreseeable conditions of use, at concentrations reaching or exceeding 0.1 ppm.
- As a minimum, specific health hazards that the employer shall address are: cancer, irritation and sensitization of the skin and respiratory system, eye and throat irritation, and acute toxicity.
- Manufacturers and importers who produce or import formaldehyde or formaldehyde-containing products shall provide downstream employers using or handling these products with an objective determination through the required labels and MSDSs if these items may constitute a health hazard within the meaning of 29 CFR 1910.1200(d) under normal conditions of use.
- The employer shall assure that hazard warning labels complying with the requirements of 29 CFR 1910.1200(f) are affixed to all containers of materials, except to the extent that 29 CFR 1910.1200(f) is inconsistent with this paragraph.
- As a minimum, for all materials capable of releasing formaldehyde at levels of 0.1 ppm to 0.5 ppm, labels shall identify that the product contains formaldehyde; list the name and address of the responsible party; and state that physical and health hazard information is readily available from the employer and material safety data sheets.
- For materials capable of releasing formaldehyde at levels above 0.5 ppm, labels shall appropriately address all hazards as defined in 29 CFR 1910.1200(d) and 29 CFR 1910.1200 Appendices A and B, including respiratory sensitization, and shall contain the words "Potential Cancer Hazard."

- ❑ In making the determinations of anticipated levels of formaldehyde release, the employer may rely on objective data indicating the extent of potential formaldehyde release under reasonably foreseeable conditions of use.

- ❑ The employer may use warning labels required by other statutes, regulations, or ordinances which impart the same information as the warning statements.

- ❑ Any employer who uses formaldehyde-containing materials shall comply with the requirements of 29 CFR 1910.1200(g) with regard to the development and updating of material safety data sheets.

- ❑ Manufacturers, importers, and distributors of formaldehyde-containing materials shall assure that MSDSs and updated information are provided to all employers purchasing such materials at the time of the initial shipment and at the time of the first shipment after a material safety data sheet is updated.

- ❑ The employer shall develop, implement, and maintain at the workplace, a written hazard communication program for formaldehyde exposures in the workplace which, at a minimum, describes how the requirements for labels and other forms of warning and MSDSs for employee information and training, will be met. Employers in multi-employer workplaces shall comply with the requirements of 29 CFR 1910.1200(e)(2).

- ❑ The employer shall assure that all employees who are assigned to workplaces where there is exposure to formaldehyde participate in a training program, except where the employer can show, using objective data, that employees are not exposed to formaldehyde at or above 0.1 ppm.

- ❑ Employers shall provide information and training to employees at the time of initial assignment, and whenever a new exposure to formaldehyde is introduced into the work area. The training shall be repeated at least annually.

- ❑ The training program shall be conducted in a manner which the employee is able to understand and shall include:

  - A discussion of the contents of this regulation and the contents of the material safety data sheet;
  - The purpose for and a description of the medical surveillance program, including:
    - A description of the potential health hazards associated with exposure to formaldehyde and a description of the signs and symptoms of exposure to formaldehyde.
    - Instructions to immediately report to the employer the development of any adverse signs or symptoms that the employee suspects is attributable to formaldehyde exposure.
  - Description of operations in the work area where formaldehyde is present and an explanation of the safe work practices appropriate for limiting exposure to formaldehyde in each job;
  - The purpose for, proper use of, and limitations of personal protective clothing and equipment;

- Instructions for the handling of spills, emergencies, and clean up procedures;
- An explanation of the importance of engineering and work-practice controls for employee protection and any necessary instruction in the use of these controls; and
- A review of emergency procedures including the specific duties or assignments of each employee in the event of an emergency.

❑ The employer shall inform all affected employees of the location of written training materials and shall make these materials readily available, without cost, to the affected employees.

❑ The employer shall provide, upon request, all training materials relating to the employee training program to the Assistant Secretary and the Director.

## Recordkeeping

❑ The employer shall establish and maintain an accurate record of all measurements taken to monitor employee exposure to formaldehyde. This record shall include:

- The date of measurement;
- The operation being monitored;
- The methods of sampling and analysis, and evidence of their accuracy and precision;
- The number, duration, time, and results of samples taken;
- The types of protective devices worn; and
- The names, job classifications, social security numbers, and exposure estimates of the employees whose exposures are represented by the actual monitoring results.

❑ Where the employer has determined that no monitoring is required, the employer shall maintain a record of the objective data relied upon to support the determination that no employee is exposed to formaldehyde at or above the action level.

❑ The employer shall establish and maintain an accurate record for each employee subject to medical surveillance. This record shall include:

- The name and social security number of the employee;
- The physician's written opinion;
- A list of any employee health complaints that may be related to formaldehyde exposure; and
- A copy of the medical examination results, including medical disease questionnaires and results of any medical tests mandated by the examining physician.

❑ The employer shall establish and maintain accurate records for employees subject to negative-pressure respirator fit testing required by this standard.

- This record shall include:
  - A copy of the protocol selected for respirator fit testing;
  - A copy of the results of any fit testing performed;
  - The size and manufacturer of the types of respirators available for selection; and
  - The date of the most recent fit testing, the name and social security number of each tested employee, and the respirator type and facepiece selected.
- The employer shall retain records for at least the following periods:
  - Exposure records and determinations shall be kept for at least 30 years.
  - Medical records shall be kept for the duration of employment plus 30 years.
  - Respirator fit-testing records shall be kept until replaced by a more recent record.
- Upon request, the employer shall make all records maintained available for examination and copying to the Assistant Secretary and the Director.
- The employer shall make employee exposure records, including estimates made from representative monitoring, available upon request for examination and copying to the subject employee, or former employee, and employee representatives in accordance with 29 CFR 1910.20(a) through (e) and (g) through (i).
- Employee medical records shall be provided, upon request, for examination and copying to the subject employee, or former employee, or to anyone having the specific written consent of the subject employee or former employee in accordance with 29 CFR 19.20(a) through (e), and (g) through (i).

# Inorganic Arsenic

## 1910.1018

Section 12 applies to all occupational exposures to inorganic arsenic except employee exposures in agriculture or resulting from pesticide application, the treatment of wood with preservatives, or the utilization of arsenically preserved wood.

### Definitions

- *Action level* means a concentration of inorganic arsenic of 5 micrograms per cubic meter of air (5 mg/$m^3$) averaged over any 8-hour period.

- *Assistant Secretary* means the Assistant Secretary of Labor for Occupational Safety and Health, U.S. Department of Labor, or designee.

- *Authorized person* means any person specifically authorized by the employer whose duties require the person to enter a regulated area, or any person entering such an area as a designated representative of employees for the purpose of exercising the right to observe monitoring and measuring procedures.

- *Director* means the Director, National Institute for Occupational Safety and Health, U.S. Department of Health and Human Services, or designee.

- *Inorganic arsenic* means copper aceto-arsenite and all inorganic compounds containing arsenic except arsine, measured as arsenic (As).

## Permissible Exposure Limit (PEL)

- ❑ The employer shall assure that no employee is exposed to inorganic arsenic at concentrations greater than 10 micrograms per cubic meter of air (10 $mg/m^3$), averaged over any 8-hour period.

## Notification of Use

- ❑ By October 1, 1978, or within 60 days after the introduction of inorganic arsenic into the workplace, every employer who is required to establish a regulated area in his workplaces shall report in writing to the OSHA area office for each such workplace:
  - The address;
  - The approximate number of employees who will be working in regulated areas; and
  - A brief summary of the operations creating the exposure and the actions which the employer intends to take to reduce exposures.
- ❑ Whenever there has been a significant change in this information, the employer shall report the changes in writing within 60 days of the changes to the OSHA area office.

## Exposure Monitoring

- ❑ Determinations of airborne exposure levels shall be made from air samples that are representative of each employee's exposure to inorganic arsenic over an 8-hour period.
- ❑ For the purposes of this section, employee exposure is that exposure which would occur if the employee were not using a respirator.
- ❑ The employer shall collect full-shift (for at least seven continuous hours) personal samples, including at least one sample for each shift for each job classification in each work area.
- ❑ Each employer who has a workplace or work operation shall monitor each such workplace and work operation to accurately determine the airborne concentration of inorganic arsenic to which employees may be exposed.
- ❑ If the initial monitoring reveals employee exposure to be below the action level, the measurements need not be repeated.
- ❑ If the initial monitoring or subsequent monitoring reveals employee exposure to be above the permissible exposure limit (PEL), the employer shall repeat monitoring at least quarterly.
- ❑ If the initial monitoring or subsequent monitoring reveals employee exposure to be above the action level and below the PEL, the employer shall repeat monitoring at least every six months.
- ❑ The employer shall continue monitoring at the required frequency until at least two consecutive measurements, taken at least seven days apart, are below the action level, at which time the employer may discontinue monitoring for that employee.

- Whenever there has been a production, process, control, or personal change which may result in new or additional exposure to inorganic arsenic, or whenever the employer has any other reason to suspect a change which may result in new or additional exposures to inorganic arsenic, additional monitoring shall be conducted.
- Within five working days after the receipt of monitoring results, the employer shall notify each employee in writing of the results that represent that employee's exposures.
- Whenever the results indicate that the representative employee exposure exceeds the PEL, the employer shall include in the written notice a statement that the PEL was exceeded and a description of the corrective action taken to reduce exposure to or below the PEL.
- The employer shall use a method of monitoring and measurement which has an accuracy, with a confidence level of 95 percent of not less than plus or minus 25 percent for concentrations of inorganic arsenic greater than or equal to 10 mg/m$^3$.
- The employer shall use a method of monitoring and measurement which has an accuracy, with a confidence level of 95 percent of not less than plus or minus 35 percent for concentrations of inorganic arsenic greater than 5 mg/m$^3$, but less than 10 mg/m$^3$.

## Regulated Areas

- The employer shall establish regulated areas where worker exposures to inorganic arsenic, without regard to the use of respirators, are in excess of the permissible limit.
- Regulated areas shall be demarcated and segregated from the rest of the workplace in any manner that minimizes the number of persons who will be exposed to inorganic arsenic.
- Access to regulated areas shall be limited to authorized persons.
- All persons entering a regulated area shall be supplied with a respirator.
- The employer shall assure that in regulated areas, food or beverages are not consumed; smoking products, chewing tobacco, and gum are not used; and cosmetics are not applied, except that these activities may be conducted in the lunchrooms, change rooms, and showers. Drinking water may be consumed in the regulated area.

## Methods of Compliance

- The employer shall institute engineering and work-practice controls to reduce exposures to or below the PEL, except to the extent that the employer can establish that such controls are not feasible.
- Where engineering and work-practice controls are not sufficient to reduce exposures to or below the PEL, they shall be used to reduce exposures to the lowest levels achievable by these controls and shall be supplemented by the use of respirators and other necessary personal

protective equipment. Employee rotation is not required as a control strategy before respiratory protection is instituted.

- ❑ The employer shall establish and implement a written program to reduce exposures to or below the permissible exposure limit by means of engineering and work-practice controls.

- ❑ Written plans for these compliance programs shall include at least the following:

  - A description of each operation in which inorganic arsenic is emitted (e.g. machinery used, material processed, controls in place, crew size, and operating procedures and maintenance practices);
  - Engineering plans and studies used to determine methods selected for controlling exposure to inorganic arsenic;
  - A report of the technology considered in meeting the PEL;
  - Monitoring data;
  - A detailed schedule for implementation of the engineering controls and work practices that cannot be implemented immediately, and for adoption and implementation of any additional engineering and work practices necessary to meet the PEL;
  - An analysis of the effectiveness of the various controls and a program to minimize the discomfort and maximize the effectiveness of respirator use whenever the employer will not achieve the PEL with engineering controls and work practices; and
  - Other relevant information.

- ❑ The employer shall install engineering controls and institute work practices on the quickest schedule feasible.

- ❑ Written plans for such a program shall be submitted upon request to the Assistant Secretary and the Director, and shall be available at the work site for examination and copying by the Assistant Secretary, Director, any affected employee, or authorized employee representatives.

- ❑ The plans required by this paragraph shall be revised and updated at least every six months to reflect the current status of the program.

## Respiratory Protection

- ❑ Respirators must be used during:

  - Periods necessary to install or implement feasible engineering or work-practice controls.
  - Work operations, such as maintenance and repair activities, for which the employer establishes that engineering and work-practice controls are not feasible.
  - Work operations for which engineering and work-practice controls are not yet sufficient to reduce employee exposures to or below the PEL.
  - Emergencies.

- ❑ The employer must implement a respiratory protection program in accordance with 29 CFR 1910.134(b) through (d) (except (d)(1)(iii)) and (f) through (m).

- If an employee exhibits breathing difficulty during fit testing or respirator use, they must be examined by a physician trained in pulmonary medicine to determine whether they can use a respirator while performing the required duty.

- The employer must use Table 4-12-1 to select the appropriate respirator or combination of respirators for inorganic arsenic compounds without significant vapor pressure, and Table 4-12-2 to select the appropriate respirator or combination of respirators for inorganic arsenic compounds that have significant vapor pressure.

- When employee exposures exceed the PEL for inorganic arsenic and also exceed the relevant limit for other gases (for example, sulfur dioxide), an air-purifying respirator provided to the employee as specified by this section must have a combination high-efficiency filter with an appropriate gas sorbent. (See footnote #1 for Table 4-12-1.)

- Employees required to use respirators may choose, and the employer must provide, a powered air-purifying respirator if it will provide proper protection. In addition, the employer must provide a combination dust and acid-gas respirator to employees who are exposed to gases over the relevant exposure limits.

## Table 4-12-1: Respiratory Protection for Inorganic Arsenic Compounds without Significant Vapor Pressure

| Concentration of inorganic arsenic (as As) or condition of use | Required respirator |
|---|---|
| Unknown, greater than 20,000 µg/m³ (20 mg/m³) or firefighting | Any full facepiece self-contained breathing apparatus operated in positive-pressure mode. |
| Not greater than 20,000 µg/m³ (20 mg/m³) | Supplied-air respirator with full facepiece, hood, helmet, or suit and operated in positive-pressure mode. |
| Not greater than 10,000 µg/m³ (10 mg/m³) | Powered air-purifying respirators in all inlet face coverings with high-efficiency filters[1]; or |
| | Half-mask[2] supplied-air respirators operated in positive-pressure mode. |
| Not greater than 500 µg/m³ | Full facepiece air-purifying respirator equipped with high-efficiency filter[1]; or |
| | Any full facepiece supplied-air respirator; or |
| | Any full facepiece self-contained breathing apparatus. |
| Not greater than 100 µg/m³ | Half-mask[2] air-purifying respirator equipped with high-efficiency filter[1]; or<br>Any half-mask[2] supplied-air respirator. |

[1] High-efficiency filter is defined as a filter with 99.97 percent efficiency against 0.3 micrometer monodisperse diethyl-hexyl phthalate (DOP) particles.
[2] Half-mask respirators shall not be used for protection against arsenic trichloride, as it is rapidly absorbed through the skin.

## Table 4-12-2: Respiratory Protection for Inorganic Arsenicals Compounds without Significant Vapor Pressure

| Concentration of inorganic arsenic (as As) or condition of use | Required respirator |
|---|---|
| Unknown, greater than 20,000 μg/m³ (20 mg/m³) or firefighting | Any full facepiece self-contained breathing apparatus operated in positive-pressure mode. |
| Not greater than 20,000 μg/m³ (20 mg/m³) | Supplied-air respirator with full facepiece, hood, helmet, or suit and operated in positive-pressure mode. |
| Not greater than 10,000 μg/m³ (10 mg/ m³) | Half-mask[2] supplied-air respirator operated in positive-pressure mode. |
| Not greater than 500 μg/m³ | Front- or back-mounted gas mask equipped with high-efficiency filter[1] and acid gas canister; or |
| | Any full facepiece supplied-air respirator; or |
| | Any full facepiece self-contained breathing apparatus. |
| Not greater than 100 μg/m³ | Half-mask[2] air-purifying respirator equipped with high-efficiency filter[1] and acid gas cartridge; or |
| | Any half-mask[2] supplied-air respirator. |

[1]High-efficiency filter is defined as a filter with 99.97 percent efficiency against 0.3 micrometer monodisperse diethyl-hexyl phthalate (DOP) particles.
[2]Half-mask respirators shall not be used for protection against arsenic trichloride, as it is rapidly absorbed through the skin.

## Protective Clothing and Equipment

- ❑ Where the possibility of skin or eye irritation from inorganic arsenic exists, and for all workers working in regulated areas, the employer shall provide, at no cost to the employee, and assure that employees use appropriate and clean protective work clothing and equipment such as, but not limited to:
  - Coveralls or similar full-body work clothing;
  - Gloves, and shoes or coverlets;
  - Face shields or vented goggles when necessary to prevent eye irritation, in accordance with the requirements of 29 CFR 1910.133(a)(2) through (6); and
  - Impervious clothing for employees subject to exposure to arsenic trichloride.
- ❑ The employer shall provide the protective clothing in a freshly laundered and dry condition at least weekly, and daily if the employee works in areas where exposures are over 100 mg/m³ of inorganic arsenic or where more frequent washing is needed to prevent skin irritation.
- ❑ The employer shall clean, launder, or dispose of protective clothing.
- ❑ The employer shall repair or replace the protective clothing and equipment as needed to maintain their effectiveness.

- The employer shall assure that all protective clothing is removed only in change rooms at the completion of a work shift.

- The employer shall assure that contaminated protective clothing that is to be cleaned, laundered, or disposed of, is placed in a closed container in the change room which prevents dispersion of inorganic arsenic outside the container.

- The employer shall inform in writing any person who cleans or launders clothing required by this section of the potentially harmful effects, including the carcinogenic effects, of exposure to inorganic arsenic.

- The employer shall assure that containers of contaminated protective clothing and equipment in the workplace, or which are to be removed from the workplace, are labeled similar to the sign at right.

CAUTION

CLOTHING CONTAMINATED WITH INORGANIC ARSENIC;

DO NOT REMOVE DUST BY BLOWING OR SHAKING.

DISPOSE OF INORGANIC ARSENIC-CONTAMINATED WASH WATER IN ACCORDANCE WITH APPLICABLE LOCAL, STATE, OR FEDERAL REGULATIONS.

- The employer shall prohibit the removal of inorganic arsenic from protective clothing or equipment by blowing or shaking.

## Housekeeping

- All surfaces shall be maintained as free as practicable of accumulations of inorganic arsenic.

- Floors and other accessible surfaces contaminated with inorganic arsenic may not be cleaned by the use of compressed air, and shoveling and brushing may be used only where vacuuming or other relevant methods have been tried and found not to be effective.

- Where vacuuming methods are selected, the vacuums shall be used and emptied in a manner to minimize the reentry of inorganic arsenic into the workplace.

- A written housekeeping and maintenance plan shall be kept which shall list appropriate frequencies for carrying out housekeeping operations, and for cleaning and maintaining dust collection equipment. The plan shall be available for inspection by the Assistant Secretary.

- Periodic cleaning of dust collection and ventilation equipment, and checks of their effectiveness, shall be carried out to maintain the effectiveness of the system and a notation kept of the last check of effectiveness and cleaning or maintenance.

## Hygiene Facilities and Practices

- The employer shall provide clean change rooms equipped with separate storage facilities for street clothes and protective clothing and equipment for employees working in regulated areas or subject to the possibility of skin or eye irritation from inorganic arsenic, in accordance with 29 CFR 1910.141(e).

- ❑ The employer shall assure that employees working in regulated areas, or subject to the possibility of skin or eye irritation from inorganic arsenic, shower at the end of the work shift.

- ❑ The employer shall provide shower facilities in accordance with 29 CFR 1910.141(d)(3).

- ❑ The employer shall provide lunchroom facilities that have a temperature controlled, positive-pressure, filtered air supply, and which are readily accessible to employees working in regulated areas.

- ❑ The employer shall assure that employees working in the regulated area, or subject to the possibility of skin or eye irritation from exposure to inorganic arsenic, wash their hands and face prior to eating.

- ❑ The employer shall provide lavatory facilities that comply with 29 CFR 1910.141(d)(1)–(2).

- ❑ For employees working in areas where exposure, without regard to the use of respirators, exceeds 100 $mg/m^3$, the employer shall provide facilities in which employees vacuum their protective clothing, and clean or change shoes before entering change rooms, lunchrooms, or shower rooms and assure that employees use such facilities.

- ❑ The employer shall assure that no employee is exposed to skin or eye contact with arsenic trichloride, or to skin or eye contact with liquid or particulate inorganic arsenic which is likely to cause skin or eye irritation.

## Medical Surveillance

- ❑ The employer shall institute a medical surveillance program for the following employees:
  - All employees who are, or will be, exposed above the action level, without regard to the use of respirators, at least 30 days per year; and
  - All employees who have been exposed above the action level, without regard to respirator use, for 30 days or more per year for a total of ten years or more of combined employment with the employer or predecessor employers prior to or after the effective date of this standard. The determination of exposures prior to the effective date of this standard shall be based upon prior exposure records, comparison with the first measurements taken, or comparison with records of exposures in areas with similar processes, extent of engineering controls utilized, and materials used by that employer.

- ❑ The employer shall assure that all medical examinations and procedures are performed by or under the supervision of a licensed physician, and shall provide these examinations without cost to the employee, without loss of pay, and at a reasonable time and place.

- ❑ For employees initially covered, or thereafter at the time of initial assignment to an area where the employee is likely to be exposed over the action level at least 30 days per year, the employer shall provide each affected employee an opportunity for a medical examination, including at least the following elements:

- A work history and a medical history which shall include a smoking history and the presence and degree of respiratory symptoms such as breathlessness, cough, sputum production and wheezing;
- A 14 x 17 inch posterior-anterior chest x-ray and International Labor Office UICC/Cincinnati (ILO U/C) rating;
- A nasal and skin examination; and
- Other examinations which the physician believes appropriate because of the employees exposure to inorganic arsenic or required respirator use.

❑ The employer shall provide the examinations at least annually for covered employees who are under 45 years of age with fewer than ten years of exposure over the action level without regard to respirator use.

❑ The employer shall provide the examinations at least semiannually, and the x-ray requirement at least annually, for other covered employees.

❑ Whenever a covered employee has not taken the examinations within six months preceding the termination of employment, the employer shall provide such examinations to the employee.

❑ If the employee for any reason develops signs or symptoms commonly associated with exposure to inorganic arsenic, the employer shall provide an appropriate examination and emergency medical treatment.

❑ The employer shall provide the following information to the examining physician:

- A copy of 29 CFR 1910.1018 and its appendices;
- A description of the affected employee's duties as they relate to the employee's exposure;
- The employee's representative exposure level or anticipated exposure level;
- A description of any personal protective equipment used, or to be used; and
- Information from previous medical examinations of the affected employee which is not readily available to the examining physician.

❑ The employer shall obtain a written opinion from the examining physician which shall include:

- The results of the medical examination and tests performed;
- The physician's opinion as to whether the employee has any detected medical conditions which would place the employee at increased risk of material impairment of the employee's health from exposure to inorganic arsenic;
- Any recommended limitations upon the employee's exposure to inorganic arsenic, or upon the use of protective clothing or equipment such as respirators; and
- A statement that the employee has been informed by the physician of the results of the medical examination and any medical conditions which require further explanation or treatment.

❑ The employer shall instruct the physician not to reveal in the written opinion specific findings or diagnoses unrelated to occupational exposure.

- ❑ The employer shall provide a copy of the written opinion to the affected employee.

## Communication of Hazards to Employees

- ❑ The employer shall institute a training program for all employees who are subject to exposure to inorganic arsenic above the action level without regard to respirator use, or for whom there is the possibility of skin or eye irritation from inorganic arsenic. The employer shall assure that those employees participate in the training program.

- ❑ The training program shall be provided for employees at the time of initial assignment and at least annually for other covered employees thereafter.

- ❑ The employer shall assure that each employee is informed of the following:
  - The quantity, location, manner of use, storage, sources of exposure, and the specific nature of operations which could result in exposure to inorganic arsenic as well as any necessary protective steps;
  - The purpose, proper use, and limitation of respirators;
  - The purpose and a description of the medical surveillance program; and
  - The engineering controls and work practices associated with the employee's job assignment.

- ❑ The employer shall make readily available to all affected employees a copy 29 CFR 1910.1018 and its appendices.

- ❑ The employer shall provide, upon request, all materials relating to the employee information and training program to the Assistant Secretary and the Director.

- ❑ The employer may use labels or signs required by other statutes, regulations, or ordinances in addition to, or in combination with required signs and labels.

- ❑ The employer shall assure that no statement appears on or near any sign or label that contradicts or detracts from the meaning of the required sign or label.

- ❑ The employer shall post signs to demarcate regulated areas [that look similar to the sample at right].

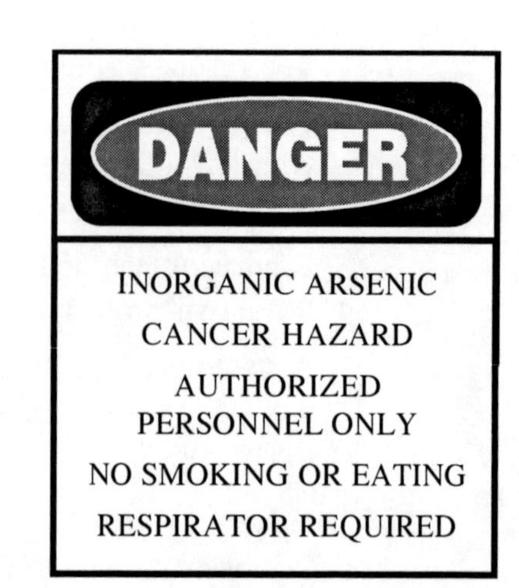


- ❑ The employer shall assure that signs are illuminated and cleaned as necessary so that the legend is readily visible.

- ❑ The employer shall apply precautionary labels to all shipping and storage containers of inorganic arsenic, and to all products containing inorganic arsenic, except when the inorganic arsenic in the product is bound in such a manner so as to make unlikely the possibility of airborne exposure to inorganic arsenic. (Possible examples of products not requiring labels are semiconductors, light emitting diodes and glass.)

The label shall look similar to the sample at right.

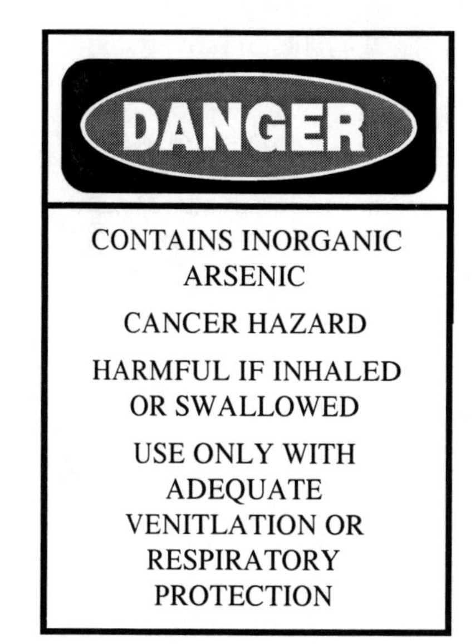


## Recordkeeping

- The employer shall establish and maintain an accurate record of all monitoring.
- This record shall include:
  - The date(s), number, duration, location, and results of each of the samples taken, including a description of the sampling procedure used to determine representative employee exposure where applicable;
  - A description of the sampling and analytical methods used, and evidence of their accuracy;
  - The type of respiratory protective devices worn, if any;
  - Name, social security number, and job classification of the employees monitored and of all other employees whose exposure the measurement is intended to represent; and
  - The environmental variables that could affect the measurement of the employee's exposure.
- The employer shall maintain these monitoring records for at least 40 years or for the duration of employment plus 20 years, whichever, is longer.
- The employer shall establish and maintain an accurate record for each employee subject to medical surveillance.
- This record shall include:
  - The name, social security number, and description of duties of the employee;
  - A copy of the physician's written opinions;
  - Results of any exposure monitoring done for that employee and the representative exposure levels supplied to the physician; and
  - Any employee medical complaints related to exposure to inorganic arsenic.
- The employer shall in addition keep, or assure that the examining physician keeps, the following medical records:
  - A copy of the medical examination results including medical and work history;
  - A description of the laboratory procedures and a copy of any standards or guidelines used to interpret the test results or references to this information;
  - The initial x-ray;
  - The x-rays for the most recent five years; and
  - Any x-rays with a demonstrated abnormality and all subsequent X-rays.
- The employer shall maintain, or assure that the physician maintains, these medical records for at least 40 years, or for the duration of employment plus 20 years, whichever is longer.

- ❑ The employer shall make available, upon request, all records required to be maintained to the Assistant Secretary and the Director for examination and copying.
- ❑ Records required by this paragraph shall be provided, upon request to employees, designated representatives, and the Assistant Secretary, in accordance with 29 CFR 1910.20(a) through (e) and (g) through (i).
- ❑ Whenever the employer ceases to do business, the successor employer shall receive and retain all records.
- ❑ Whenever the employer ceases to do business and there is no successor employer to receive and retain the records for the prescribed period, these records shall be transmitted to the Director.
- ❑ At the expiration of the retention period for the records required to be maintained, the employer shall notify the Director at least three months prior to the disposal of such records and shall transmit those records to the Director if he requests them within that period.
- ❑ The employer shall also comply with any additional requirements involving the transfer of records set in 29 CFR 1910.20(h).

## Observation of Monitoring

- ❑ The employer shall provide affected employees or their designated representatives an opportunity to observe any monitoring of employee exposure to inorganic arsenic conducted.
- ❑ Whenever observation of the monitoring of employee exposure to inorganic arsenic requires entry into an area where the use of respirators, protective clothing, or equipment is required, the employer shall provide the observer with and assure the use of such respirators, clothing, and such equipment, and shall require the observer to comply with all other applicable safety and health procedures.
- ❑ Without interfering with the monitoring, observers shall be entitled to:
  - Receive an explanation of the measurement procedures;
  - Observe all steps related to the monitoring of inorganic arsenic performed at the place of exposure; and
  - Record the results obtained or receive copies of the results when returned by the laboratory.

Part IV — Section 13

# Lead

## 1910.1025

> Section 13 applies to occupational exposure to lead. This section does not apply to the construction industry or to certain agricultural operations.

## Definitions

- *Action level* means employee exposure, without regard to the use of respirators, to an airborne concentration of lead of 30 micrograms per cubic meter of air (30 mg/m$^3$) averaged over an 8-hour period.
- *Assistant Secretary* means the Assistant Secretary of Labor for Occupational Safety and Health, U.S. Department of Labor, or designee.
- *Chelation* means using an external agent (chemical) to reduce the amounts of lead in the body.
- *Director* means the Director, National Institute for Occupational Safety and Health (NIOSH), U.S. Department of Health and Human Services, or designee.
- *Lead* means metallic lead, all inorganic lead compounds, and organic lead soaps. Excluded from this definition are all other organic lead compounds.

## Permissible Exposure Limit (PEL)

- The employer shall assure that no employee is exposed to lead at concentrations greater than 50 micrograms per cubic meter of air (50 mg/m$^3$) averaged over an 8-hour period.

- If an employee is exposed to lead for more than eight hours in any work day, the PEL, as a time-weighted average (TWA) for that day, shall be reduced according to the following formula:

**Maximum permissible limit (in mg/m$^3$) = 400 • hours worked in the day**

- When respirators are used to supplement engineering and work-practice controls, employee exposure, for the purpose of determining whether the employer has complied with the PEL, may be considered to be at the level provided by the protection factor of the respirator for those periods the respirator is worn. Those periods may be averaged with exposure levels during periods when respirators are not worn to determine the employee's daily TWA exposure.

## Exposure Monitoring

- Employee exposure is that exposure which would occur if the employee were not using a respirator.

- The employer shall collect full-shift (for at least seven continuous hours) personal samples, including at least one sample for each shift for each job classification in each work area.

- Full-shift personal samples shall be representative of the monitored employee's regular, daily exposure to lead.

- Each employer shall determine if any exployee may be exposed to lead at or above the action level.

- The employer shall monitor employee exposures and base initial determinations on the employee exposure monitoring results and any of the following, relevant considerations:
    - Monitoring for the initial determination which may be limited to a representative sample of the exposed employees who the employer reasonably believes are exposed to the greatest airborne concentrations of lead in the workplace.
    - Measurements of airborne lead made in the preceding 12 months, if the sampling and analytical methods used meet the accuracy and confidence levels.

- Where a determination shows the possibility of any employee exposure at or above the action level, the employer shall conduct monitoring which is representative of the exposure for each employee in the workplace who is exposed to lead.

- Where a determination is made that no employee is exposed to airborne concentrations of lead at or above the action level, the employer shall make a written record of such determination. The record shall include the date of determination, location within the worksite, and the name and social security number of each employee monitored.

- If the initial monitoring reveals employee exposure to be below the action level, the measurements need not be repeated.

- If the initial determination or subsequent monitoring reveals employee exposure to be at or above the action level, but below the PEL, the employer shall repeat monitoring at least every six months. The employer shall continue monitoring at the required frequency until at least two consecutive measurements, taken at least seven days apart, are below the action level, at which time the employer may discontinue monitoring for that employee.

- If the initial monitoring reveals that employee exposure is above the PEL, the employer shall repeat monitoring quarterly. The employer shall continue monitoring at the required frequency until at least two consecutive measurements, taken at least seven days apart, are below the PEL but at or above the action level, at which time the employer shall repeat monitoring for that employee.

- Whenever there has been a production, process, control, or personnel change which may result in new or additional exposure to lead, or whenever the employer has any other reason to suspect a change which may result in new or additional exposures to lead, additional monitoring shall be conducted.

- Within five working days after the receipt of monitoring results, the employer shall notify each employee in writing of the results that represent that employee's exposure.

- Whenever the results indicate that the representative employee's exposure, without regard to respirators, exceeds the PEL, the employer shall include in the written notice a statement that the PEL was exceeded and a description of the corrective action taken, or to be taken, to reduce exposure to or below the PEL.

- The employer shall use a method of monitoring and analysis which has an accuracy, to a confidence level of 95 percent, of not less than plus or minus 20 percent for airborne concentrations of lead equal to or greater than 30 $mg/m^3$.

## Methods of Compliance

- Where any employee is exposed to lead above the PEL for more than 30 days per year, the employer shall implement engineering and work-practice controls (including administrative controls) to reduce and maintain employee exposure to lead in accordance with the implementation schedule in Table 4-13-1, except to the extent that the employer can demonstrate that such controls are not feasible.

- Wherever the engineering and work-practice controls which can be instituted are not sufficient to reduce employee exposure to or below the permissible exposure limit, the employer shall use them to reduce exposures to the lowest feasible level and supplement them by the use of respiratory protection.

- Where any employee is exposed to lead above the PEL, but for 30 days or less per year, the employer shall implement engineering controls to reduce exposures to 200 $mg/m^3$, but thereafter may implement any combination of engineering, work-practice (including administrative controls), and respiratory controls to reduce and maintain employee exposure to lead to or below 50 $mg/m^3$.

## Table 4-13-1: Implementation Schedule

| Industry | Compliance Dates: (50 μg/m³)[1] |
|---|---|
| Lead chemicals, secondary copper smelting | July 19, 1996 |
| Nonferrous foundries | July 19, 1996[2] |
| Brass and bronze ingot manufacture | 6 years[3] |

[1] Calculated by counting from the date the stay on implementation of paragraph (e)(1) was lifted by the U.S. Court of Appeals for the District of Columbia, the number of years specified in the 1978 lead standard, and subsequent amendments for compliance with the PEL of 50 μ g/m³ for exposure to airborne concentrations of lead levels for the particular industry.
[2] Large nonferrous foundries (20 or more employees) are required to achieve the PEL of 50 μg/m³ by means of engineering and work-practice controls. Small nonferrous foundries (fewer than 20 employees) are required to achieve an 8-hour TWA of 75 μg/m³ by such controls.
[3] Expressed as the number of years from the date on which the Court lifts the stay on the implementation of paragraph (e)(1) for this industry for employers to achieve a lead in air concentration of 75 μg/m³. Compliance with paragraph (e) in this industry is determined by a compliance directive that incorporates elements from the settlement agreement between OSHA and representatives of the industry.

## Written Program

- ❑ Each employer shall establish and implement a written compliance program to reduce exposures to or below the PEL and interim levels, if applicable, solely by means of engineering and work-practice controls.
- ❑ Written plans for these compliance programs shall include at least the following:
  - A description of each operation in which lead is emitted (e.g. machinery used, material processed, controls in place, crew size, employee job responsibilities, operating procedures, and maintenance practices);
  - A description of the specific means that will be employed to achieve compliance, including engineering plans and studies used to determine methods selected for controlling exposure to lead;
  - A report of the technology considered in meeting the PEL;
  - Air monitoring data which documents the source of lead emissions;
  - A detailed schedule for implementation of the program, including documentation such as copies of purchase orders for equipment, construction contracts, etc.;
  - A work-practice program;
  - An administrative control schedule; and
  - Other relevant information.
- ❑ Written programs shall be submitted upon request to the Assistant Secretary and the Director, and shall be available at the worksite for examination and copying by the Assistant Secretary, Director, any affected employee, and authorized employee representatives.

- Written programs shall be revised and updated at least every six months to reflect the current status of the program.

## Mechanical Ventilation

- When ventilation is used to control exposure, measurements that demonstrate the effectiveness of the system in controlling exposure, such as capture velocity, duct velocity, or static pressure, shall be made at least every three months.

- Measurements of the system's effectiveness in controlling exposure shall be made within five days of any change in production, process, or control that might result in a change in employee exposure to lead.

- If air from exhaust ventilation is recirculated into the workplace, the employer shall assure:
  - The system has a high-efficiency filter with reliable back-up filter; and
  - Controls are installed, operating, and maintained to monitor the concentration of lead in the return air and to bypass the recirculation system automatically if it fails.

## Administrative Controls

- If administrative controls are used as a means of reducing employees' TWA exposures to lead, the employer shall establish and implement a job rotation schedule which includes:
  - Name or identification number of each affected employee;
  - Duration and exposure levels at each job or work station where each affected employee is located; and
  - Any other information which may be useful in assessing the reliability of administrative controls to reduce exposure to lead.

## Respiratory Protection

- Where engineering and work-practice controls do not reduce employee exposure to or below the 50 $mg/m^3$ PEL, the employer shall supplement these controls with respirators.

- Respirators must be used during:
  - Periods necessary to install or implement engineering or work-practice controls.
  - Work operations for which engineering and work-practice controls are not sufficient to reduce employee exposures to or below the PEL.
  - Periods when an employee requests a respirator.

- The employer must implement a respiratory protection program in accordance with 29 CFR 1910.134(b) through (d) (except (d)(1)(iii)) and (f) through (m).

- If an employee has breathing difficulty during fit testing or respirator use, the employer must provide the employee with a medical examination to determine whether or not the employee can use a respirator while performing the required duty.

- The employer must select the appropriate respirator or combination of respirators from Table 4-13-2.

- The employer must provide a powered air-purifying respirator instead of the respirator specified in Table 4-13-2 of this section when an employee chooses to use this type of respirator and such a respirator provides adequate protection to the employee.

## Table 4-13-2: Respiratory Protection for Lead Aerosols

| Airborne concentration of lead or condition of use | Required respirator[1] |
|---|---|
| Not in excess of 0.5 mg/m³ (10 X PEL) | Half-mask, air-purifying respirator equipped with high-efficiency filters. |
| Not in excess of 2.5 mg/m³ (50 X PEL) | Full facepiece, air-purifying respirator with high-efficiency filters.[3] |
| Not in excess of 50 mg/m³ (1,000 X PEL) | Any powered, air-purifying respirator with high-efficiency filters[3]; or<br>Half-mask, supplied-air respirator operated in positive-pressure mode.[2] |
| Not in excess of 100 mg/m³ (2,000 X PEL) | Supplied-air respirators with full facepiece, hood, helmet, or suit, operated in positive-pressure mode. |
| Greater than 100 mg/m³, unknown concentration, or firefighting | Full facepiece, self-contained breathing apparatus operated in positive-pressure mode. |

[1] Respirators specified for high concentrations can be used at lower concentrations of lead.
[2] Full facepiece is required if the lead aerosols cause eye or skin irritation at the use concentrations.
[3] A high-efficiency particulate filter means 99.97 percent efficient against 0.3 micron size particles.

## Protective Clothing and Equipment

- If an employee is exposed to lead above the PEL, without regard to the use of respirators or where the possibility of skin or eye irritation exists, the employer shall provide at no cost to the employee, and assure that the employee uses, appropriate protective work clothing and equipment such as, but not limited to:
  - Coveralls or similar full-body work clothing;
  - Gloves, hats, and shoes or disposable shoe coverlets; and
  - Face shields, vented goggles, or other appropriate protective equipment which complies with 29 CFR 1910.133.

- The employer shall provide the protective clothing in a clean and dry condition at least weekly, and daily to employees whose exposure levels without regard to a respirator are over 200 mg/m$^3$ of lead as an 8-hour TWA.

- The employer shall provide for the cleaning, laundering, or disposal of protective clothing and equipment.

- The employer shall repair or replace required protective clothing and equipment as needed to maintain their effectiveness.

- The employer shall assure that all protective clothing is removed only in change rooms at the completion of a work shift.

- The employer shall assure that contaminated protective clothing which is to be cleaned, laundered, or disposed of, is placed in a closed container in change rooms which prevents dispersion of lead outside the container.

- The employer shall inform in writing any person who cleans or launders protective clothing or equipment of the potentially harmful effects of exposure to lead.

- The employer shall assure that the containers of contaminated protective clothing and equipment are labeled similar to the sample at right.

- The employer shall prohibit the removal of lead from protective clothing or equipment by blowing, shaking, or any other means that disperses lead into the air.

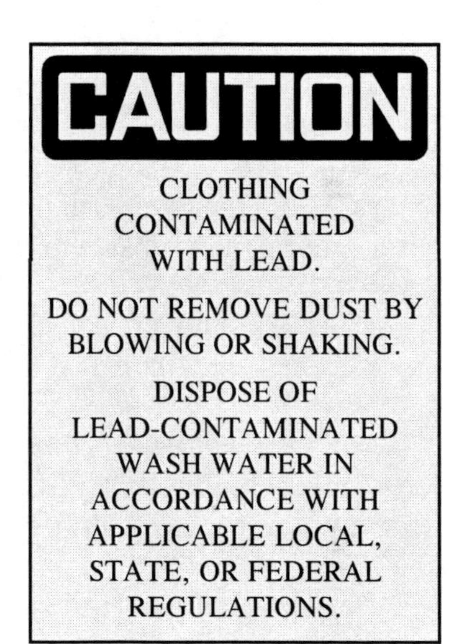


## Housekeeping

- All surfaces shall be maintained as free as practicable of accumulations of lead.

- Floors and other surfaces where lead accumulates may not be cleaned by the use of compressed air.

- Shoveling, dry or wet sweeping, and brushing may be used only where vacuuming or other equally effective methods have been tried and found not to be effective.

- Where vacuuming methods are selected, the vacuums shall be used and emptied in a manner that minimizes the reentry of lead into the workplace.

## Hygiene Facilities and Practices

- The employer shall assure that in areas where employees are exposed to lead above the PEL, without regard to the use of respirators, food or beverages are not present or consumed, tobacco products are not present or used, and cosmetics are not applied, except in change rooms, lunchrooms, and showers.

- ❑ The employer shall provide clean change rooms for employees who work in areas where their airborne exposure to lead is above the PEL, without regard to the use of respirators.
- ❑ The employer shall assure that change rooms are equipped with separate storage facilities for protective work clothing and equipment and street clothes which prevent cross-contamination.
- ❑ The employer shall assure that employees who work in areas where their airborne exposure to lead is above the PEL, without regard to the use of respirators, shower at the end of the work shift.
- ❑ The employer shall provide shower facilities in accordance with 29 CFR 1910.141(d)(3).
- ❑ The employer shall assure that employees who are required to shower do not leave the workplace wearing any clothing or equipment worn during the work shift.
- ❑ The employer shall provide lunchroom facilities for employees who work in areas where their airborne exposure to lead is above the PEL, without regard to the use of respirators.
- ❑ The employer shall assure that lunchroom facilities have a temperature controlled, positive-pressure, filtered air supply, and are readily accessible to employees.
- ❑ The employer shall assure that employees who work in areas where their airborne exposure to lead is above the PEL, without regard to the use of a respirator, wash their hands and face prior to eating, drinking, smoking, or applying cosmetics.
- ❑ The employer shall assure that employees do not enter lunchroom facilities with protective work clothing or equipment unless surface lead dust has been removed by vacuuming, downdraft booth, or other cleaning method.
- ❑ The employer shall provide an adequate number of lavatory facilities that comply with 29 CFR 1910.141(d)(1) and (2).

## Medical Surveillance

- ❑ The employer shall institute a medical surveillance program for all employees who are, or may be, exposed above the action level for more than 30 days per year.
- ❑ The employer shall assure that all medical examinations and procedures are performed by or under the supervision of a licensed physician.
- ❑ The employer shall provide the required medical surveillance, including multiple physician review, without cost to employees, and at a reasonable time and place.

- The employer shall make available biological monitoring in the form of blood sampling and analysis for lead and zinc protoporphyrin (ZPP) levels to each employee on the following schedule:

  - At least every six months to each employee.
  - At least every two months for each employee whose last blood sampling and analysis indicated a blood lead level at or above 40 mg/100 g of whole blood. This frequency shall continue until two consecutive blood samples and analyses indicate a blood lead level below 40 mg/100 g of whole blood.
  - At least monthly during the removal period of each employee removed from exposure to lead due to an elevated blood lead level.

- Whenever the results of a blood lead level test indicate that an employee's blood lead level exceeds the numerical criterion for medical removal, the employer shall provide a second (follow-up) blood sampling test within two weeks after the employer receives the results of the first blood sampling test.

- Blood lead level sampling and analysis provided pursuant to this section shall have an accuracy (to a confidence level of 95 percent) of plus or minus 15 percent or 6 mg/100 ml, whichever is greater, and shall be conducted by a laboratory licensed by the Center for Disease Control, (CDC) or a laboratory which has received a satisfactory grade in blood lead proficiency testing from the CDC in the prior 12 months.

- Within five working days after the receipt of biological monitoring results, the employer shall notify in writing each employee whose blood lead level exceeds 40 mg/100 g of the employee's blood lead level and the standard's requirement of temporary medical removal with medical removal protection benefits when an employee's blood lead level exceeds the numerical criterion for medical removal.

- The employer shall make available medical examinations and consultations to each employee covered on the following schedule:

  - At least annually for each employee for whom a blood sampling test conducted at any time during the preceding 12 months indicated a blood lead level at or above 40 m g/100 g.
  - Prior to each employee being assigned for the first time to an area in which airborne concentrations of lead are at or above the action level.
  - As soon as possible upon notification by an employee either that the employee has developed signs or symptoms commonly associated with lead intoxication, the employee desires medical advice concerning the effects of current or past exposure to lead on the employee's ability to procreate a healthy child, or the employee has demonstrated difficulty in breathing during a respirator fitting test or during use.
  - As medically appropriate for each employee either removed from exposure to lead due to a risk of sustaining material impairment to health, or otherwise limited pursuant to a final medical determination.

- ❑ Medical examinations shall include the following elements:
  - A detailed work history and medical history with particular attention to past lead exposure (occupational and non-occupational), personal habits (smoking, hygiene), and past gastrointestinal, hematologic, renal, cardiovascular, reproductive, and neurological problems;
  - A thorough physical examination with particular attention to teeth, gums, and hematologic, gastrointestinal, renal, cardiovascular, and neurological systems;
  - Pulmonary status should be evaluated if respiratory protection will be used;
  - A blood pressure measurement;
  - A blood sample and analysis which determines:
    - Blood lead level;
    - Hemoglobin and hematocrit determinations, red cell indices, and examination of peripheral smear morphology;
    - Zinc protoporphyrin;
    - Blood urea nitrogen; and
    - Serum creatinine.
  - A routine urinalysis with microscopic examination; and
  - Any laboratory or other test which the examining physician deems necessary by sound medical practice.
- ❑ The content of medical examinations shall be determined by an examining physician and, if requested by an employee, shall include pregnancy testing or laboratory evaluation of male fertility.
- ❑ If the employer selects the initial physician who conducts any medical examination or consultation provided to an employee, the employee may designate a second physician:
  - To review any findings, determinations, or recommendations of the initial physician; and
  - To conduct such examinations, consultations, and laboratory tests as the second physician deems necessary to facilitate this review.
- ❑ The employer shall promptly notify an employee of the right to seek a second medical opinion after each occasion that an initial physician conducts a medical examination or consultation pursuant to this section. The employer may condition its participation in, and payment for, the multiple physician review mechanism upon the employee doing the following within 15 days after receipt of the foregoing notification, or receipt of the initial physician's written opinion, whichever is later:
  - The employee informing the employer that he or she intends to seek a second medical opinion, and
  - The employee initiating steps to make an appointment with a second physician.
- ❑ If the findings, determinations, or recommendations of the second physician differ from those of the initial physician, then the employer and the employee shall assure that efforts are made for the two physicians to resolve any disagreement.

- If the two physicians have been unable to quickly resolve their disagreement, then the employer and the employee, through their respective physicians, shall designate a third physician:
  - To review any findings, determinations, or recommendations of the prior physicians; and
  - To conduct such examinations, consultations, laboratory tests, and discussions with the two prior physicians as the third physician deems necessary to resolve the disagreement of the physicians.
- The employer shall act consistent with the findings, determinations, and recommendations of the third physician, unless the employer and the employee reach an agreement that is otherwise consistent with the recommendations of at least one of the three physicians.
- The employer shall provide an initial physician conducting a medical examination or consultation under this section with the following information:
  - A copy of 29 CFR 1910.1025 including all appendices;
  - A description of the affected employee's duties as they relate to the employee's exposure;
  - The employee's exposure level or anticipated exposure level to lead and any other toxic substance (if applicable);
  - A description of any personal protective equipment used or to be used;
  - Prior blood lead determinations; and
  - All prior written medical opinions concerning the employee in the employer's possession or control.
- The employer shall provide the foregoing information to a second or third physician conducting a medical examination or consultation under this section upon request either by the second or third physician, or by the employee.
- The employer shall obtain and furnish the employee with a copy of a written medical opinion from each examining or consulting physician which contains the following information:
  - The physician's opinion as to whether the employee has any detected medical condition which would place the employee at increased risk of material impairment of the employee's health from exposure to lead;
  - Any recommended special protective measures to be provided to the employee, or limitations to be placed upon the employee's exposure to lead;
  - Any recommended limitation upon the employee's use of respirators, including a determination of whether the employee can wear a powered air-purifying respirator if a physician determines that the employee cannot wear a negative-pressure respirator; and
  - The results of the blood lead determinations.
- The employer shall instruct each examining and consulting physician to:
  - Not reveal either in the written opinion or in any other means of communication to the employer, findings, including laboratory results, or diagnoses unrelated to an employee's occupational exposure to lead; and

- Advise the employee of any medical condition, occupational or non-occupational, which dictates further medical examination or treatment.

- ❑ The employer and an employee, or authorized employee representative, may agree upon the use of any expeditious alternate physician determination mechanism in lieu of the multiple physician review mechanism as long as the same requirements are met.

- ❑ The employer shall assure that any person whom he retains, employs, supervises, or controls, does not engage in prophylactic chelation of any employee at any time.

- ❑ If therapeutic or diagnostic chelation is to be performed, the employer shall assure that it be done under the supervision of a licensed physician in a clinical setting with thorough and appropriate medical monitoring and that the employee is notified in writing prior to its occurrence.

## Medical Removal Protection

- ❑ The employer shall remove an employee from work having an exposure to lead at or above the action level on each occasion that a periodic and a follow-up blood sampling test indicate that the employee's blood lead level is at or above 60 mg/100 g of whole blood.

- ❑ The employer shall remove an employee from work having an exposure to lead at or above the action level on each occasion that the average of the last three blood sampling tests conducted pursuant to this section, or the average of all blood sampling tests conducted over the previous 6 months, whichever is longer, indicates that the employee's blood lead level is at or above 50 mg/100 g of whole blood; provided, however, that an employee need not be removed if the last blood sampling test indicates a blood lead level at or below 40 mg/100 g of whole blood.

- ❑ The employer shall remove an employee from work having an exposure to lead at or above the action level on each occasion that a final medical determination results in a medical finding, determination, or opinion that the employee has a detected medical condition which places the employee at increased risk of material impairment to health from exposure to lead.

- ❑ The phrase "final medical determination" shall mean the outcome of the multiple physician review mechanism or alternate medical determination mechanism used pursuant to the medical surveillance.

- ❑ Where a final medical determination results in any recommended special protective measures for an employee, or limitations on an employee's exposure to lead, the employer shall implement and act consistently with the recommendation.

- ❑ The employer shall return an employee to his or her former job status:

  - For an employee removed due to a blood lead level at or above 60 mg/100 g, or due to an average blood lead level at or above 50 mg/100 g, when two consecutive blood sampling tests indicate that the employee's blood lead level is at or below 40 mg/100 g of whole blood.

  - For an employee removed due to a final medical determination, when a subsequent final medical determination results in a medical finding, determination, or opinion that the employee no longer has a detected medical condition which places the employee at increased risk of material impairment to health from exposure to lead.

- The requirement that an employer return an employee to his or her former job status is not intended to expand upon or restrict any rights an employee has or would have had, absent temporary medical removal, to a specific job classification or position under the terms of a collective bargaining agreement.

- The employer shall remove any limitations placed on an employee, or end any special protective measures provided to an employee pursuant to a final medical determination when a subsequent final medical determination indicates that the limitations or special protective measures are no longer necessary.

- Where the multiple physician review mechanism, or alternate medical determination mechanism used pursuant to the medical surveillance provisions of this section, has not yet resulted in a final medical determination with respect to an employee, the employer shall act as follows:

  - The employer may remove the employee from exposure to lead, provide special protective measures to the employee, or place limitations upon the employee, consistent with the medical findings, determinations, or recommendations of any of the physicians who have reviewed the employee's health status.
  - The employer may return the employee to his or her former job status, end any special protective measures provided to the employee, and remove any limitations placed upon the employee, consistent with the medical findings, determinations, or recommendations of any of the physicians who have reviewed the employee's health status, with two exceptions:
    - If the initial removal, special protection, or limitation of the employee resulted from a final medical determination which differed from the findings, determinations, or recommendations of the initial physician, then the employer shall await a final medical determination.
    - If the employee has been on removal status for the preceding 18 months due to an elevated blood lead level, then the employer shall await a final medical determination.

- The employer shall provide to an employee up to 18 months of medical removal protection benefits on each occasion that an employee is removed from exposure to lead or otherwise limited.

- The requirement that an employer provide medical removal protection benefits means that the employer shall maintain the earnings, seniority, and other employment rights and benefits of an employee as though the employee had not been removed from normal exposure to lead or otherwise limited.

- During the period of time that an employee is removed from normal exposure to lead or otherwise limited, the employer may condition the provision of medical removal protection benefits upon the employee's participation in follow-up medical surveillance.

- ❑ Where an employer removes an employee from exposure to lead or otherwise places limitations on an employee due to the effects of lead exposure on the employee's medical condition, the employer shall provide medical removal protection benefits to the employee.

- ❑ If a removed employee files a claim for workers' compensation payments for a lead-related disability, then the employer shall continue to provide medical removal protection benefits pending disposition of the claim. To the extent that an award is made to the employee for earnings lost during the period of removal, the employer's medical removal protection obligation shall be reduced by such amount. The employer shall receive no credit for workers' compensation payments received by the employee for treatment-related expenses.

- ❑ The employer's obligation to provide medical removal protection benefits to a removed employee shall be reduced to the extent that the employee receives compensation for earnings lost during the period of removal either from a publicly- or employer-funded compensation program, or receives income from employment with another employer made possible by virtue of the employee's removal.

- ❑ The employer shall take the following measures with respect to any employee removed from exposure to lead due to an elevated blood lead level whose blood lead level has not declined within the past 18 months of removal so that the employee has been returned to his or her former job status:

  - The employer shall make available to the employee a medical examination to obtain a final medical determination with respect to the employee.
  - The employer shall assure that the final medical determination obtained indicates whether or not the employee may be returned to his or her former job status, and if not, what steps should be taken to protect the employee's health.
  - Where the final medical determination has not yet been obtained, or once obtained indicates that the employee may not yet be returned to his or her former job status, the employer shall continue to provide medical removal protection benefits to the employee until either the employee is returned to former job status, or a final medical determination is made that the employee is incapable of ever safely returning to his or her former job status.
  - Where the employer acts pursuant to a final medical determination which permits the return of the employee to his or her former job status despite what would otherwise be an unacceptable blood lead level, later questions concerning removing the employee again shall be decided by a final medical determination. The employer need not automatically remove such an employee pursuant to the blood lead level removal criteria.

## Medical Recordkeeping

- ❑ The employer shall establish and maintain an accurate record for each employee subject to medical surveillance.

- ❑ This record shall include:

  - The name, social security number, and description of the duties of the employee;

- A copy of the physician's written opinions;
- Results of any airborne exposure monitoring done for that employee and the representative exposure levels supplied to the physician; and
- Any employee medical complaints related to exposure to lead.

- The employer shall keep, or assure that the examining physician keeps, the following medical records:
  - A copy of the medical examination results, including medical and work history;
  - A description of the laboratory procedures and a copy of any standards or guidelines used to interpret the test results or references to that information; and
  - A copy of the results of biological monitoring.

- The employer shall maintain or assure that the physician maintains those medical records for at least 40 years or for the duration of employment plus 20 years, whichever is longer.

- The employer shall establish and maintain an accurate record for each employee removed from current exposure to lead.

- Each record shall include:
  - The name and social security number of the employee;
  - The date, on each occasion, that the employee was removed from current exposure to lead as well as the corresponding date on which the employee was returned to his or her former job status;
  - A brief explanation of how each removal was, or is being, accomplished; and
  - A statement with respect to each removal indicating whether or not the reason for the removal was an elevated blood lead level.

- The employer shall maintain each medical removal record for at least the duration of an employee's employment.

- The employer shall make available, upon request, all records required to be maintained to the Assistant Secretary and the Director for examination and copying.

- Environmental monitoring, medical removal, and medical records shall be provided, upon request, to employees, designated representatives, and the Assistant Secretary, in accordance with 29 CFR 1910.20(a) through (e) and (2)(i). Medical removal records shall be provided in the same manner as environmental monitoring records.

- Whenever the employer ceases to do business, the successor employer shall receive and retain all records.

- Whenever the employer ceases to do business and there is no successor employer to receive and retain the records, these records shall be transmitted to the Director.

- ❑ At the expiration of the retention period for the records required to be maintained by this section, the employer shall notify the Director at least three months prior to the disposal of such records and shall transmit those records to the Director if requested within the period.

- ❑ The employer shall also comply with any additional requirements involving transfer of records set forth in 29 CFR 1910.20(h).

## Communication of Hazard to Employees

- ❑ Each employer who has a workplace in which there is a potential exposure to airborne lead at any level shall inform the employees.

- ❑ The employer shall institute a training program for, and assure the participation of, all employees who are subject to exposure to lead at or above the action level, or for whom the possibility of skin or eye irritation exists.

- ❑ The employer shall provide initial training for those employees at the time of initial job assignment.

- ❑ The training program shall be repeated at least annually for each employee.

- ❑ The employer shall assure that each employee is informed of the following:
    - The content of 29 CFR 1910.1025 and appendices;
    - The specific nature of the operations which could result in exposure to lead above the action level;
    - The purpose, proper selection, fitting, use, and limitations of respirators;
    - The purpose and description of the medical surveillance program and the medical removal protection program, including information concerning the adverse health effects associated with excessive exposure to lead (with particular attention to the adverse reproductive effects on both males and females);
    - The engineering controls and work practices associated with the employee's job assignment;
    - The contents of any compliance plan in effect; and
    - Instructions to employees that chelating agents should not routinely be used to remove lead from their bodies, and should not be used at all except under the direction of a licensed physician.

- ❑ The employer shall make readily available to all affected employees a copy of 29 CFR 1910.1025 and its appendices.

- ❑ The employer shall provide, upon request, all materials relating to the employee information and training program to the Assistant Secretary and the Director.

- ❑ The employer shall include as part of the training program, and shall distribute to employees, any materials pertaining to the Occupational Safety and Health Act, and the regulations issued pursuant to that Act which are made available to the employer by the Assistant Secretary.

- The employer shall assure that no statement appears on or near any sign that contradicts or detracts from the meaning of the required sign.
- In work areas where the PEL is exceeded, the employer shall post warning signs that look similar to the sample at right.
- The employer shall assure that signs required by this paragraph are illuminated and cleaned as necessary so that the legend is readily visible.

## Recordkeeping

- The employer shall establish and maintain an accurate record of all monitoring.
- This record shall include:
  - The date(s), number, duration, location, and results of each of the samples taken, including a description of the sampling procedure used to determine representative employee exposure, where applicable;
  - A description of the sampling and analytical methods used, and evidence of their accuracy;
  - The type of respiratory protective devices worn, if any;
  - Name, social security number, and job classification of the employee monitored and of all other employees whose exposure the measurement is intended to represent; and
  - The environmental variables that could affect the measurement of employee exposure.
- The employer shall maintain these monitoring records for at least 40 years or for the duration of employment plus 20 years, whichever is longer.

## Observation of Monitoring

- The employer shall provide affected employees or their designated representatives an opportunity to observe any monitoring of employee exposure to lead.
- Whenever observation of the monitoring of employee exposure to lead requires entry into an area where the use of respirators or protective clothing and equipment is required, the employer shall provide the observer with, and assure the use of, such respirators, clothing, and equipment, and shall require the observer to comply with all other applicable safety and health procedures.
- Without interfering with the monitoring, observers shall be entitled to:
  - Receive an explanation of the measurement procedures;
  - Observe all steps related to the monitoring of lead performed at the place of exposure; and
  - Record the results obtained or receive copies of the results from the laboratory.

# Methylene Chloride

## 1910.1052

29 CFR 1910.1052 establishes requirements for employers to control occupational exposure to methylene chloride (MC). Employees exposed to MC are at increased risk of developing cancer; adverse effects on the heart, central nervous system, and liver; and skin or eye irritation. Exposure may occur through inhalation, absorption through the skin, or contact with the skin. MC is a solvent that is used in many different types of work activities, such as paint stripping, polyurethane foam manufacturing, and cleaning and degreasing. Each covered employer must make an initial determination of each employee's exposure to MC. If the employer determines that employees are exposed below the action level, the only other provisions that apply are that a record must be made of the determination, the employees must receive information and training and, where appropriate, employees must be protected from contact with liquid MC. Section 14 applies to all occupational exposures to MC, CAS No. 75-09-2, in general industry, construction, and shipyard employment.

### Definitions

- *Action level* means a concentration of airborne MC of 12.5 parts per million (12.5 ppm) calculated as an 8-hour time-weighted average (TWA).

- *Assistant Secretary* means the Assistant Secretary of Labor for Occupational Safety and Health, U.S. Department of Labor, or designee.

- *Authorized person* means any person specifically authorized by the employer and required by work duties to be present in regulated areas, or any person entering such an area as a designated representative of employees for the purpose of exercising the right to observe monitoring and measuring procedures.

- *Director* means the Director of the National Institute for Occupational Safety and Health, U.S. Department of Health and Human Services, or designee.

- *Emergency* means any occurrence, such as, but not limited to, equipment failure, rupture of containers, or failure of control equipment, which results, or is likely to result, in an uncontrolled release of MC. If an incidental release of MC can be controlled by employees such as maintenance personnel at the time of release and in accordance with the leak/spill provisions, it is not considered an emergency.

- *Employee exposure* means exposure to airborne MC that occurs or would occur if the employee were not using respiratory protection.

- *Methylene chloride* (MC) means an organic compound with chemical formula, $CH_2C_{12}$. Its CAS No. is 75-09-2. Its molecular weight is 84.9 g/mole.

- *Physician* or other licensed health care professional is an individual whose legally permitted scope of practice (i.e., license, registration, or certification) allows him or her to independently provide, or be delegated, the responsibility to provide some or all of the health care services.

- *Regulated area* means an area, demarcated by the employer, where an employee's exposure to airborne concentrations of MC exceeds or can reasonably be expected to exceed either the 8-hour TWA PEL or STEL.

- *STEL* means short-term exposure limit as determined by any 15-minute sample period.

- *Symptom* means central nervous system effects such as headaches, disorientation, dizziness, fatigue, and decreased attention span; skin effects such as chapping, erythema, cracked skin, or skin burns; and cardiac effects such as chest pain or shortness of breath.

## Permissible Exposure Limit (PEL)

- The employer shall ensure that no employee is exposed to an airborne concentration of MC in excess of 25 parts of MC per million parts of air (25 ppm) as an 8-hour TWA.

- The employer shall ensure that no employee is exposed to an airborne concentration of MC in excess of 125 parts of MC per million parts of air (125 ppm) as determined over a sampling period of 15 minutes.

## Exposure Monitoring

- Where MC is present in the workplace, the employer shall determine each employee's exposure by either:
  - Taking a personal breathing zone air sample of each employee's exposure; or
  - Taking personal breathing zone air samples that are representative of each employee's exposure.

- The employer may consider personal breathing zone air samples to be representative of employee exposures when they are taken as follows:

- The employer has taken one or more personal breathing zone air samples for at least one employee in each job classification in a work area during every work shift, and the employee sampled is expected to have the highest MC exposure.
- The employer has taken one or more personal breathing zone air samples which indicate the highest likely 15-minute exposures during such operations for at least one employee in each job classification in the work area during every work shift, and the employee sampled is expected to have the highest MC exposure.
- Personal breathing zone air samples taken during one work shift may be used to represent employee exposures on other work shifts where the employer can document that the tasks performed and conditions in the workplace are similar across shifts.

❑ The employer shall ensure that the methods used to perform exposure monitoring produce results that are accurate to a confidence level of 95 percent, and are:

- Within plus or minus 25 percent for airborne concentrations of MC above the 8-hour TWA PEL or STEL; or
- Within plus or minus 35 percent for airborne concentrations of MC at or above the action level but at or below the 8-hour TWA PEL.

❑ Each employer whose employees are exposed to MC shall perform initial exposure monitoring to determine each affected employee's exposure, except under the following conditions:

- Where objective data demonstrate that MC cannot be released in the workplace in airborne concentrations at or above the action level or above the STEL. The objective data shall represent the highest MC exposures likely to occur under reasonably foreseeable conditions of processing, use, or handling. The employer shall document the objective data exemption.
- Where the employer has performed exposure monitoring under conditions substantially equivalent to existing conditions.
- Where employees are exposed to MC on fewer than 30 days per year (e.g., on a construction site), and the employer has measurements by direct-reading instruments which give immediate results (such as a detector tube) and provide sufficient information regarding employee exposures to determine what control measures are necessary to reduce exposures to acceptable levels.
- Where the initial determination shows employee exposures at or above the action level or above the STEL. The employer shall establish an exposure monitoring program for periodic monitoring of employee exposure to MC in accordance with Table 4-14-1.

## Table 4-14-1: Six Initial Determination Exposure Scenarios and Their Associated Monitoring Frequencies

| Exposure scenario | Required monitoring activity |
|---|---|
| Below the action level and at or below the STEL | No 8-hour TWA or STEL monitoring required. |
| Below the action level and above the STEL | No 8-hour TWA monitoring required; monitor STEL exposures every three months. |
| At or above the action level, at or below the TWA, and at or below the STEL | Monitor 8-hour TWA exposures every six months. |
| At or above the action level, at or below the TWA, and above the STEL | Monitor 8-hour TWA exposures every six months and monitor STEL exposures every three months. |
| Above the TWA and at or below the STEL | Monitor 8-hour TWA exposures every three months. |
| Above the TWA and above the STEL | Monitor 8-hour TWA exposures and STEL exposures every three months. |

Note: The employer may decrease the frequency of exposure monitoring to every six months when at least two consecutive measurements taken at least seven days apart show exposures to be at or below the 8-hour TWA PEL. The employer may discontinue the periodic 8-hour TWA monitoring for employees where at least two consecutive measurements taken at least seven days apart are below the action level. The employer may discontinue the periodic STEL monitoring for employees where at least two consecutive measurements taken at least seven days apart are at or below the STEL.

- ❑ The employer shall perform exposure monitoring when a change in workplace conditions indicates that employee exposure may have increased. Examples of situations that may require additional monitoring include changes in production, process, control equipment, or work practices, or a leak, rupture, or other breakdown.

- ❑ Where exposure monitoring is performed due to a spill, leak, rupture, or equipment breakdown, the employer shall clean up the MC and perform the appropriate repairs before monitoring.

- ❑ The employer shall, within 15 working days after the receipt of the results of any monitoring performed under this section, notify each affected employee of these results in writing, either individually or by posting results in an appropriate location that is accessible to affected employees.

- ❑ Whenever monitoring results indicate that employee exposure is above the 8-hour TWA PEL or STEL, the employer shall describe in the written notification the corrective action being taken to reduce employee exposure to or below the 8-hour TWA PEL or STEL and the schedule for completion of this action.

- ❑ The employer shall provide affected employees or their designated representatives an opportunity to observe any monitoring of employee exposure to MC.

- ❑ When observation of the monitoring of employee exposure to MC requires entry into an area where the use of protective clothing or equipment is required, the employer shall provide, at no cost to the observer(s), and the observer(s) shall be required to use such clothing and equipment and shall comply with all other applicable safety and health procedures.

## Regulated Areas

- The employer shall establish a regulated area wherever an employee's exposure to airborne concentrations of MC exceeds or can reasonably be expected to exceed either the 8-hour TWA PEL or STEL.
- The employer shall limit access to regulated areas to authorized persons.
- The employer shall supply a respirator to each person who enters a regulated area and shall require each affected employee to use that respirator whenever MC exposures are likely to exceed the 8-hour TWA PEL or STEL.
- An employer who has implemented all feasible engineering, work-practice, and administrative controls, and who has established a regulated area where MC exposure can be reliably predicted to exceed the 8-hour TWA PEL or STEL only on certain days (for example, because of work or process schedule) would need to have affected employees use respirators in that regulated area only on those days.
- The employer shall ensure that, within a regulated area, employees do not engage in non-work activities that may increase dermal or oral MC exposure.
- The employer shall ensure that while employees are wearing respirators, they do not engage in activities (such as taking medication or chewing gum or tobacco) which interfere with respirator seal or performance.
- The employer shall demarcate regulated areas from the rest of the workplace in any manner that adequately establishes and alerts employees to the boundaries of the area and minimizes the number of authorized employees exposed to MC within the regulated area.
- An employer at a multi-employer worksite who establishes a regulated area shall communicate the access restrictions and locations of these areas to all other employers with work operations at that worksite.

## Methods of Compliance

- The employer shall institute and maintain the effectiveness of engineering controls and work practices to reduce employee exposure to or below the PELs except to the extent that the employer can demonstrate that such controls are not feasible.
- Wherever the feasible engineering controls and work practices that can be instituted are not sufficient to reduce employee exposure to or below the 8-hour TWA PEL or STEL, the employer shall use them to reduce employee exposure to the lowest levels achievable by these controls and shall supplement them by the use of respiratory protection.

- ❑ The employer shall not implement a schedule of employee rotation as a means of compliance with the PELs.

- ❑ The employer shall implement procedures to detect leaks of MC in the workplace. In work areas where spills may occur, the employer shall make provisions to contain any spills and to safely dispose of any MC-contaminated waste materials.

- ❑ The employer shall ensure that all incidental leaks are repaired and that all incidental spills are cleaned promptly by employees who use the appropriate personal protective equipment and are trained in proper methods of cleanup.

## Respiratory Protection

- ❑ Respirators must be used during:
  - Periods when an employee's exposure to MC exceeds the 8-hour TWA PEL or STEL (for example, when an employee is using MC in a regulated area).
  - Periods necessary to install or implement feasible engineering and work-practice controls.
  - A few work operations, such as some maintenance operations and repair activities, for which the employer demonstrates that engineering and work-practice controls are infeasible.
  - Work operations for which feasible engineering and work-practice controls are not sufficient to reduce employee exposures to or below the PELs.
  - Emergencies.

- ❑ The employer must implement a respiratory protection program in accordance with 29 CFR 1910.134(b) through (m) (except (d)(1)(iii), and (d)(3)(iii)(B)(1) and (2)).

- ❑ Employers who provide employees with gas masks with organic-vapor canisters for the purpose of emergency escape must replace the canisters after any emergency use and before the gas masks are returned to service.

- ❑ The employer must select appropriate atmosphere-supplying respirators from Table 4-14-2.

## Table 4-14-2: Respiratory Protection for Airborne Methylene Chloride

| Methylene chloride airborne concentration (ppm) or condition of use | Respirator required[1] |
|---|---|
| Up to 625 ppm (25 X PEL) | Continuous flow supplied-air respirator, hood, or helmet. |
| Up to 1,250 ppm (50 X 8-hour TWA PEL) | Full facepiece supplied-air respirator operated in negative-pressure (demand) mode; or<br>Full facepiece self-contained breathing apparatus (SCBA) operated in negative-pressure (demand) mode. |
| Up to 5,000 ppm (200 X 8-hour TWA PEL) | Continuous flow supplied-air respirator, full facepiece; or<br>Pressure-demand supplied-air respirator, full facepiece; or<br>Positive-pressure full facepiece SCBA. |
| Unknown concentration, or above 5,000 ppm (greater than 200 X 8-hour TWA PEL) | Positive-pressure full facepiece SCBA; or<br>Full facepiece pressure-demand supplied-air respirator with an auxiliary self-contained air supply. |
| Firefighting | Positive-pressure full facepiece SCBA. |
| Emergency escape | Any continuous flow or pressure-demand SCBA.<br>Gas mask with organic vapor canister. |

[1] Respirators assigned for higher airborne concentrations may be used at lower concentrations.

- The employer shall provide a respirator at no cost to each affected employee, and ensure that each affected employee uses such respirator where appropriate. Respirators shall be used during:
    - When an employee's exposure to MC exceeds, or can reasonably be expected to exceed, the 8-hour TWA PEL or STEL (such as when an employee is using MC in a regulated area).
    - The time interval necessary to install or implement feasible engineering and work-practice controls.
    - Work operations, such as some maintenance operations and repair activities, for which the employer demonstrates that engineering and work-practice controls are infeasible.
    - Where feasible engineering and work-practice controls are not sufficient to reduce exposures to or below the PELs.
    - Emergencies.
- Before having any employee use a supplied-air respirator in the negative-pressure mode, or a gas mask with organic vapor canister for emergency escape, the employer shall have a physician or other licensed health care professional ascertain each affected employee's ability to use such respiratory protection. The physician or other licensed health care professional shall provide his or her findings to the affected employee and the employer in a written opinion.
- The appropriate atmosphere-supplying respirators, as specified in Table 4-14-2, shall be selected from those approved by the National Institute for Occupational Safety and Health (NIOSH) under the provisions of 42 CFR Part 84, "Respiratory Protective Devices." When

employers elect to provide gas masks with organic vapor canisters for use in emergency escape, the organic vapor canisters shall bear the approval of NIOSH.

- ❑ The employer shall permit employees who wear respirators to leave the regulated area to readjust the facepieces to their faces to achieve a proper fit, and to wash their faces and respirator facepieces as necessary in order to prevent skin irritation associated with respirator use.
- ❑ The employer shall ensure that each respirator issued to the employee is properly fitted and exhibits the least possible facepiece leakage from among the facepieces tested.
- ❑ The employer shall perform qualitative or quantitative fit tests at the time of initial fitting and at least annually thereafter for each employee wearing a negative-pressure respirator, including those employees for whom emergency escape respirators are provided.

*Note: The only supplied-air respirators to which this provision would apply are SCBA in negative-pressure mode and full facepiece supplied-air respirators operated in negative-pressure mode. The small business compliance guides will contain examples of protocols for qualitative and quantitative fit testing.*

## Protective Clothing and Equipment

- ❑ Where needed to prevent MC-induced skin or eye irritation, the employer shall provide, at no cost to the employee, clean protective clothing and equipment that is resistant to MC, and shall ensure that each affected employee uses it. Eye and face protection shall meet the requirements of 29 CFR 1910.133 or 29 CFR 1915.153, as applicable.
- ❑ The employer shall clean, launder, repair, and replace all protective clothing and equipment as needed to maintain their effectiveness.
- ❑ The employer shall be responsible for the safe disposal of such clothing and equipment.

## Hygiene Facilities and Practices

- ❑ If it is reasonably foreseeable that an employee's skin may contact solutions containing 0.1 percent or greater MC (for example, through splashes, spills, or improper work practices), the employer shall provide conveniently located washing facilities capable of removing the MC, and shall ensure that affected employees use these facilities as needed.
- ❑ If it is reasonably foreseeable that an employee's eyes may contact solutions containing 0.1 percent or greater MC (for example through splashes, spills, or improper work practices), the employer shall provide appropriate eyewash facilities within the immediate work area for emergency use, and shall ensure that affected employees use those facilities when necessary.

## Medical Surveillance

- ❑ The employer shall make medical surveillance available for employees who are or may be exposed to MC:

- At or above the action level on 30 or more days per year, or above the 8-hour TWA PEL or STEL on ten or more days per year;
- Above the 8-hour TWA PEL or STEL for any time period where an employee has been identified by a physician or other licensed health care professional as being at risk from cardiac disease or from some other serious MC-related health condition and such employee requests inclusion in the medical surveillance program; and
- During an emergency.

- The employer shall provide at no cost all required medical surveillance to affected employees, without loss of pay and at a reasonable time and place.

- The employer shall ensure that a physician or other licensed health care professional performs all medical surveillance procedures.

- The employer shall make medical surveillance available to each affected employee as follows:
    - The employer shall provide initial medical surveillance under the schedule or before the time of initial assignment of the employee, whichever is later.
    - The employer shall update the medical and work history for each affected employee annually. The employer shall provide periodic physical examinations, including appropriate laboratory surveillance, as follows:
        - For employees 45 years of age or older, within 12 months of the initial surveillance or any subsequent medical surveillance; and
        - For employees younger than 45 years of age, within 36 months of the initial surveillance or any subsequent medical surveillance.

- When an employee leaves the employer's workplace, or is reassigned to an area where exposure to MC is consistently at or below the action level and STEL, medical surveillance shall be made available if six months or more have elapsed since the last medical surveillance.

- The employer shall provide additional medical surveillance at frequencies other than those listed above when recommended in the written medical opinion. (For example, the physician or other licensed health care professional may determine an examination is warranted in less than 36 months for employees younger than 45 years of age based upon evaluation of the results of the annual medical and work history.)

- The comprehensive medical and work history shall emphasize neurological symptoms, skin conditions, history of hematologic or liver disease, signs or symptoms suggestive of heart disease (angina, coronary artery disease), risk factors for cardiac disease, MC exposures, and work practices and personal protective equipment used during such exposures.

- Where physical examinations are provided, the physician or other licensed health care professional shall accord particular attention to the lungs, cardiovascular system (including blood pressure and pulse), liver, nervous system, and skin. The physician or other licensed health care professional shall determine the extent and nature of the physical examination based on the health status of the employee and analysis of the medical and work history.

- ❑ The physician or other licensed health care professional shall determine the extent of any required laboratory surveillance based on the employee's observed health status and the medical and work history.

- ❑ The medical surveillance shall also include any other information or reports the physician or other licensed health care professional determines are necessary to assess the employee's health in relation to MC exposure.

- ❑ The employer shall ensure that medical surveillance made available when an employee has been exposed to MC in emergency situations includes, at a minimum:
  - Appropriate emergency treatment and decontamination of the exposed employee;
  - A comprehensive physical examination with special emphasis on the nervous system, cardiovascular system, lungs, liver, and skin, including blood pressure and pulse;
  - An updated medical and work history, as appropriate for the medical condition of the employee; and
  - Laboratory surveillance, as indicated by the employee's health status.

- ❑ Where the physician or other licensed health care professional determines it is necessary, the scope of the medical examination shall be expanded and the appropriate additional medical surveillance, such as referrals for consultation or examination, shall be provided.

- ❑ The employer shall provide the following information to a physician or other licensed health care professional who is involved in the diagnosis of MC-induced health effects:
  - A copy of 29 CFR 1910.1052 including its applicable appendices;
  - A description of the affected employee's past, current, and anticipated future duties as they relate to the employee's MC exposure;
  - The employee's former or current exposure levels or, for employees not yet occupationally exposed to MC, the employee's anticipated exposure levels, and the frequency and exposure levels anticipated to be associated with emergencies;
  - A description of any personal protective equipment, such as respirators, used or to be used; and
  - Information from previous employment-related medical surveillance of the affected employee which is not otherwise available to the physician or other licensed health care professional.

- ❑ For each physical examination required by this section, the employer shall ensure that the physician or other licensed health care professional provides to the employer and to the affected employee a written opinion regarding the results of that examination within 15 days of completion of the evaluation of medical and laboratory findings, but not more than 30 days after the examination. The written medical opinion shall be limited to the following information:
  - The physician's or other licensed health care professional's opinion concerning whether the employee has any detected medical condition(s) which would place the employee's health at increased risk of material impairment from exposure to MC;

- Any recommended limitations upon the employee's exposure to MC or use of protective clothing or equipment and respirators;
- A statement that the employee has been informed by the physician or other licensed health care professional that MC is a potential occupational carcinogen, of risk factors for heart disease, and the potential for exacerbation of underlying heart disease by exposure to MC through its metabolism to carbon monoxide; and
- A statement that the employee has been informed by the physician or other licensed health care professional of the results of the medical examination and any medical conditions resulting from MC exposure which require further explanation or treatment.

❑ The employer shall instruct the physician or other licensed health care professional not to reveal to the employer, orally or in the written opinion, any specific records, findings, and diagnoses that have no bearing on occupational exposure to MC. The written medical opinion may also include information and opinions generated to comply with other OSHA health standards.

## Communication of Hazards to Employees

❑ The employer shall communicate the following hazards associated with MC on labels and in material safety data sheets in accordance with the requirements of the Hazard Communication Standard, 29 CFR 1910.1200, 29 CFR 1915.1200, or 29 CFR 1926.59, as appropiate: cancer, cardiac effects (including elevation of carboxyhemoglobin), central nervous system effects, liver effects, and skin and eye irritation.

❑ The employer shall provide information and training for each affected employee prior to or at the time of initial assignment to a job involving potential exposure to MC.

❑ The employer shall ensure that information and training is presented in a manner that is understandable to the employees.

❑ The employer shall train each affected employee as required under the Hazard Communication Standard, 29 CFR 1910.1200, 29 CFR 1915.1200, or 29 CFR 1926.59, as appropriate.

❑ In addition to the information required under the Hazard Communication Standard, 29 CFR 1910.1200, 29 CFR 1915.1200, or 29 CFR 1926.59, as appropriate; wherever an employee's exposure to airborne concentrations of MC exceeds, or can reasonably be expected to exceed the action level, the employer shall inform each affected employee of the quantity, location, manner of use, release, and storage of MC and the specific operations in the workplace that could result in exposure to MC, particularly noting where exposures may be above the 8-hour TWA PEL or STEL.

❑ The employer shall retrain each affected employee as necessary to ensure that each employee exposed above the action level or the STEL maintains the requisite understanding of the principles of safe use and handling of MC in the workplace.

❑ Whenever there are workplace changes, such as modifications of tasks or procedures or the institution of new tasks or procedures, which increase employee exposure, and where those exposures exceed or can reasonably be expected to exceed, the action level; the employer shall update the training as necessary to ensure that each affected employee has the requisite proficiency.

- ❑ An employer whose employees are exposed to MC at a multi-employer worksite shall notify the other employers with work operations at that site in accordance with the requirements of the Hazard Communication Standard, 29 CFR 1910.1200, 29 CFR 1915.1200, or 29 CFR 1926.59, as appropriate.

- ❑ The employer shall provide, upon request, to the Assistant Secretary or the Director all available materials relating to employee information and training.

## Recordkeeping

- ❑ Where an employer seeks to demonstrate that initial monitoring is unnecessary through reasonable reliance on objective data showing that any materials in the workplace containing MC will not release MC at levels which exceed the action level or the STEL under foreseeable conditions of exposure, the employer shall establish and maintain an accurate record of the objective data relied upon in support of the exemption.

- ❑ This record shall include at least the following information:
  - The MC-containing material in question;
  - The source of the objective data;
  - The testing protocol, results of testing, and/or analysis of the material for the release of MC;
  - A description of the operation exempted under paragraph (d)(2)(i) of this section and how the data support the exemption; and
  - Other data relevant to the operations, materials, processing, or employee exposures covered by the exemption.

- ❑ The employer shall maintain this record for the duration of the employer's reliance upon such objective data.

## Exposure Measurements Recordkeeping

- ❑ The employer shall establish and keep an accurate record of all measurements taken to monitor employee exposure to MC.

- ❑ Where the employer has 20 or more employees, this record shall include at least the following information:
  - The date of measurement for each sample taken;
  - The operation involving exposure to MC which is being monitored;
  - Sampling and analytical methods used, and evidence of their accuracy;

  - Number, duration, and results of samples taken;
  - Type of personal protective equipment, such as respiratory protective devices, worn, if any; and
  - Name, social security number, job classification, and exposure of all of the employees represented by monitoring, indicating which employees were actually monitored.

- Where the employer has fewer than 20 employees, the record shall include at least the following information:
  - The date of measurement for each sample taken;
  - Number, duration, and results of samples taken; and
  - Name, social security number, job classification, and exposure of all of the employees represented by monitoring, indicating which employees were actually monitored.

- The employer shall maintain this record for at least 30 years, in accordance with 29 CFR 1910.1020.

- The employer shall establish and maintain an accurate record for each employee subject to medical surveillance.

- The record shall include at least the following information:
  - The name, social security number, and description of the duties of the employee;
  - Written medical opinions; and
  - Any employee medical conditions related to exposure to MC.

- The employer shall ensure that this record is maintained for the duration of employment plus 30 years, in accordance with 29 CFR 1910.1020.

- The employer, upon written request, shall make all records available to the Assistant Secretary and the Director for examination and copying, in accordance with 29 CFR 1910.1020.

- The employer, upon request, shall make any employee exposure and objective data records available for examination and copying by affected employees, former employees, and designated representatives, in accordance with 29 CFR 1910.1020.

- The employer, upon request, shall make employee medical records available for examination and copying by the subject employee and anyone having the specific written consent of the subject employee, in accordance with 29 CFR 1910.1020.

- The employer shall comply with the requirements concerning transfer of records set forth in 29 CFR 1910.1020(h).

# Methylenedianiline

## 1910.1050

Section 15 applies to all occupational exposures to Methylenedianiline (MDA), CAS No. 101-77-9.

## Definitions

- *Action level* means a concentration of airborne MDA of 5 parts per billion (5 ppb) as an 8-hour time-weighted average (TWA).
- *Assistant Secretary* means the Assistant Secretary of Labor for Occupational Safety and Health, U.S. Department of Labor, or designee.
- *Authorized person* means any person specifically authorized by the employer whose duties require the person to enter a regulated area, or any person entering such an area, as a designated representative of employees, for the purpose of exercising the right to observe monitoring and measuring procedures.
- *Container* means any barrel, bottle, can, cylinder, drum, reaction vessel, storage tank, commercial packaging, or the like, but does not include piping systems.
- *Dermal exposure to MDA* occurs where employees are engaged in the handling, application, or use of mixtures or materials containing MDA with any of the following non-airborne forms of MDA:
    - Liquid, powdered, granular, or flaked mixtures containing MDA in concentrations greater than 0.1 percent by weight or volume; and
    - Materials other than "finished articles" containing MDA in concentrations greater than 0.1 percent by weight or volume.

- *Director* means the Director of the National Institute for Occupational Safety and Health, U.S. Department of Health and Human Services, or designee.

- *Emergency* means any occurrence such as, but not limited to, equipment failure, rupture of containers, or failure of control equipment, which results in an unexpected and potentially hazardous release of MDA.

- *Employee exposure* means exposure to MDA which would occur if the employee were not using a respirator or protective work clothing and equipment.

- *Finished article containing MDA* is defined as a manufactured item which:
  - Is formed to a specific shape or design during manufacture;
  - Has end use function(s) dependent in whole or part upon its shape or design during end use; and
  - Is, where applicable, fully cured by virtue of having been subjected to the conditions (such as time or temperature) necessary to complete the desired chemical reaction.

- *4,4'Methylenedianiline* (MDA) means the chemical 4,4'-diaminodiphenylmethane, CAS No. 101-77-9, in the form of a vapor, liquid, or solid. The definition also includes the salts of MDA.

- *Regulated areas* means areas where airborne concentrations of MDA exceed, or can reasonably be expected to exceed, the permissible exposure limits, or where dermal exposure to MDA can occur.

## Permissible Exposure Limit (PEL)

- The employer shall assure that no employee is exposed to an airborne concentration of MDA in excess of 10 parts per billion (ppb) as an 8-hour TWA.

- The employer shall assure that no employee is exposed to an airborne concentration of MDA in excess of 100 parts of MDA per billion (ppb) as determined over a sampling period of 15 minutes.

## Emergency Situations

- A written plan for emergency situations shall be developed for each workplace where there is a possibility of an emergency. Appropriate portions of the plan shall be implemented in the event of an emergency.

- The plan shall specifically provide that employees engaged in correcting emergency conditions shall be equipped with the appropriate personal protective equipment and clothing until the emergency is abated.

- The plan shall specifically include provisions for alerting and evacuating affected employees as well as the elements prescribed in 29 CFR 1910.38, "Employee emergency plans and fire prevention plans."

- Where there is the possibility of employee exposure to MDA due to an emergency, means shall be developed to alert promptly those employees who have the potential to be directly exposed. Affected employees not engaged in correcting emergency conditions shall be evacuated immediately in the event of an emergency. Means shall also be developed and implemented for alerting other employees who may be exposed as a result of the emergency.

## Exposure Monitoring

- Determinations of employee exposure shall be made from breathing zone air samples that are representative of each employee's exposure to airborne MDA over an 8-hour period. Determination of employee exposure to the STEL shall be made from breathing zone air samples collected over a 15-minute sampling period.

- Representative employee exposure shall be determined on the basis of one or more samples representing full-shift exposure for each shift for each job classification in each work area where exposure to MDA may occur.

- Where the employer can document that exposure levels are equivalent for similar operations in different work shifts, the employer shall only be required to determine representative employee exposure for that operation during one shift.

- Each employer who has a workplace or work operation covered by this standard shall perform initial monitoring to accurately determine the airborne concentrations of MDA to which employees may be exposed.

- If the monitoring reveals employee exposure at or above the action level, but at or below the PELs, the employer shall repeat such representative monitoring for each such employee at least every six months.

- If the monitoring reveals employee exposure above the PELs, the employer shall repeat such monitoring for each such employee at least every three months.

- The employer may alter the monitoring schedule from every three months to every six months for any employee for whom two consecutive measurements taken at least seven days apart indicate that the employee exposure has decreased to below the TWA but above the action level.

- If the initial monitoring reveals employee exposure to be below the action level, the employer may discontinue the monitoring for that employee.

- If the periodic monitoring reveals that employee exposures, as indicated by at least two consecutive measurements taken at least seven days apart, are below the action level, the employer may discontinue the monitoring for that employee.

- ❑ The employer shall institute the exposure monitoring when there has been a change in production process, chemicals present, control equipment, personnel, or work practices which may result in new or additional exposures to MDA, or when the employer has any reason to suspect a change which may result in new or additional exposures.

- ❑ Monitoring shall be accurate, to a confidence level of 95 percent, to within plus or minus 25 percent for airborne concentrations of MDA.

- ❑ The employer shall, within 15 working days after the receipt of the results of any monitoring performed under this standard, notify each employee of these results, in writing, either individually or by posting the results in an appropriate location that is accessible to affected employees.

- ❑ The written notification shall contain the corrective action being taken by the employer to reduce the employee exposure to or below the PELs, wherever the PELs are exceeded.

- ❑ The employer shall make routine inspections of employee's hands, face, and forearms potentially exposed to MDA. Other potential dermal exposures reported by the employee must be referred to the appropriate medical personnel for observation. If the employer determines that an employee has been exposed to MDA, the employer shall:
  - Determine the source of exposure;
  - Implement protective measures to correct the hazard; and
  - Maintain records of the corrective actions.

## Regulated Areas

- ❑ The employer shall establish regulated areas where airborne concentrations of MDA exceed, or can reasonably be expected to exceed, the PEL.

- ❑ Where employees are subject to dermal exposure to MDA, the employer shall establish those work areas as regulated areas.

- ❑ Regulated areas shall be demarcated from the rest of the workplace in a manner that minimizes the number of persons potentially exposed.

- ❑ Access to regulated areas shall be limited to authorized persons.

- ❑ Each person entering a regulated area shall be supplied with, and required to use, the appropriate personal protective clothing and equipment.

- ❑ The employer shall ensure that employees do not eat, drink, smoke, chew tobacco or gum, or apply cosmetics in regulated areas.

## Methods of Compliance

- The employer shall institute engineering controls and work practices to reduce and maintain employee exposure to MDA at or below the PELs except to the extent that the employer can establish that these controls are not feasible.

- Wherever the feasible engineering controls and work practices which can be instituted are not sufficient to reduce employee exposure to or below the PELs, the employer shall use them to reduce employee exposure to the lowest levels achievable by these controls, and shall supplement them by the use of respiratory protective devices.

- The employer shall establish and implement a written program to reduce employee exposure to or below the PELs by means of engineering and work-practice controls, and by use of respiratory protection where permitted under this section.

- The program shall include a schedule for periodic maintenance (e.g., leak detection) and shall include the written plan for emergency situations.

- This written program shall be furnished, upon request, for examination and copying to the Assistant Secretary, the Director, affected employees, and designated employee representatives. The employer shall review and, as necessary, update such plans at least once every 12 months to make certain they reflect the current status of the program.

- Employee rotation shall not be permitted as a means of reducing exposure.

## Respiratory Protection

- Respirators must be used during:
  - Periods necessary to install or implement feasible engineering and work-practice controls.
  - Work operations for which the employer establishes that engineering and work-practice controls are not feasible.
  - Work operations for which feasible engineering and work-practice controls are not yet sufficient to reduce employee exposure to or below the PEL.
  - Emergencies.

- The employer must implement a respiratory protection program in accordance with 29 CFR 1910.134(b) through (d) (except (d)(1)(iii)) and (f) through (m).

- The employer must select, and ensure that employees use, the appropriate respirator from Table 4-15-1.

- Any employee who cannot use a negative-pressure respirator must be given the option of using a positive-pressure respirator or a supplied-air respirator operated in the continuous-flow or pressure-demand mode.

## Table 4-15-1: Respiratory Protection for MDA

| Airborne concentration of MDA or condition of use | Respirator type[1] |
|---|---|
| Less than or equal to 10 x PEL | Half-mask respirator with HEPA[2] cartridge.[3] |
| Less than or equal to 50 x PEL | Full facepiece respirator with HEPA[2] cartridge[3] or canister. |
| Less than or equal to 1,000 x PEL | Full facepiece powered air-purifying respirator with HEPA[2] cartridges.[3] |
| Greater than 1,000 x PEL or unknown | Self-contained breathing apparatus with full facepiece in positive-pressure mode; or<br>Full facepiece positive-pressure demand supplied-air respirator with auxiliary self-contained air supply. |
| Escape | Any full facepiece air-purifying respirator with HEPA[2] cartridges,[3] or<br>Any positive-pressure or continuous flow self-contained breathing apparatus with full facepiece or hood. |
| Firefighting | Full facepiece self-contained breathing apparatus in positive-pressure demand mode. |

Notes:

[1] Respirators assigned for higher environmental concentrations may be used at lower concentrations.
[2] High efficiency particulate in air filter (HEPA) means a filter that is at least 99.97 percent efficient against mono-dispersed particles of 0.3 micrometers or larger.
[3] Combination HEPA/Organic Vapor Cartridges shall be used whenever MDA in liquid form or a process requiring heat is used.

## Protective Clothing and Equipment

- ❑ Where employees are subject to dermal exposure to MDA, liquids containing MDA can be splashed into the eyes, or airborne concentrations of MDA are in excess of the PEL, the employer shall provide, at no cost to the employee, and ensure that the employee uses, appropriate protective work clothing and equipment which prevent contact with MDA such as, but not limited to:
    - Aprons, coveralls, or other full-body work clothing;
    - Gloves, head coverings, and foot coverings;
    - Face shields or chemical goggles; and
    - Other appropriate protective equipment which comply with 29 CFR 1910.133.
- ❑ The employer shall ensure that, at the end of their work shift, employees remove MDA-contaminated protective work clothing and equipment that is not routinely removed throughout the day in provided, change rooms in accordance with the provisions established for change rooms.
- ❑ The employer shall ensure that employees remove all other MDA-contaminated protective work clothing or equipment before leaving a regulated area.

- The employer shall ensure that no employee takes MDA-contaminated work clothing or equipment out of the change room, except those employees authorized to do so for the purpose of laundering, maintenance, or disposal.

- MDA-contaminated work clothing or equipment shall be placed and stored in closed containers that prevent dispersion of the MDA outside the container.

- Containers of MDA-contaminated protective work clothing or equipment which are to be taken out of change rooms or the workplace for cleaning, maintenance, or disposal, shall bear labels warning of the hazards of MDA.

- The employer shall provide the employee with clean protective clothing and equipment. The employer shall ensure that protective work clothing or equipment required by this paragraph is cleaned, laundered, repaired, or replaced at intervals appropriate to maintain its effectiveness.

- The employer shall prohibit the removal of MDA from protective work clothing or equipment by blowing, shaking, or any methods that allow MDA to reenter the workplace.

- The employer shall ensure that laundering of MDA-contaminated clothing is done so as to prevent the release of MDA in the workplace.

- Any employer who gives MDA-contaminated clothing to another person for laundering shall inform such person of the requirement to prevent the release of MDA.

- The employer shall inform any person who launders or cleans protective clothing or equipment contaminated with MDA of the potentially harmful effects of exposure.

- MDA-contaminated clothing shall be transported in properly labeled and sealed impermeable bags or containers.

## Hygiene Facilities and Practices

- The employer shall provide clean change rooms for employees who must wear protective clothing or use protective equipment because of their exposure to MDA.

- Change rooms must be equipped with separate storage for protective clothing and equipment, and for street clothes which prevents MDA contamination of street clothes.

- The employer shall ensure that employees, who work in areas where there is the potential for exposure resulting from airborne MDA (e.g., particulates or vapors) above the action level, shower at the end of the work shift.

- Shower facilities shall comply with 29 CFR 1910.141(d)(3).

- The employer shall ensure that employees who are required to shower pursuant to the provisions contained herein do not leave the workplace wearing any protective clothing or equipment worn during the work shift.

- Where dermal exposure to MDA occurs, the employer shall ensure that materials spilled or deposited on the skin are removed as soon as possible by methods which do not facilitate the dermal absorption of MDA.

- Whenever food or beverages are consumed at the worksite and employees are exposed to MDA at or above the PEL or are subject to dermal exposure to MDA, the employer shall provide readily accessible lunch areas.

- Lunch areas located within the workplace and in areas where there is the potential for airborne exposure to MDA at or above the PEL shall have a positive-pressure, temperature controlled, filtered air supply.

- Lunch areas may not be located in areas within the workplace where the potential for dermal exposure to MDA exists.

- The employer shall ensure that employees who have been subjected to dermal exposure to MDA, or exposed to MDA above the PEL, wash their hands and faces with soap and water prior to eating, drinking, smoking, or applying cosmetics.

- The employer shall ensure that employees exposed to MDA do not enter lunch facilities with MDA-contaminated protective work clothing or equipment.

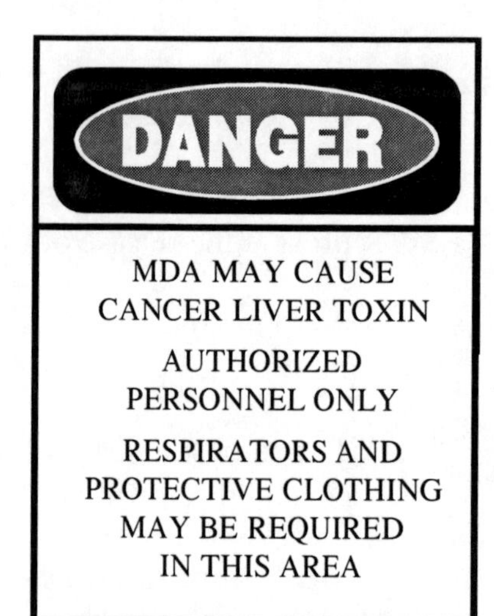


## Communication of Hazards to Employees

- The employer shall post and maintain legible signs demarcating regulated areas and entrances or access ways to regulated areas, that look similar to the sample at above right.

- The employer shall ensure that labels or other appropriate warnings are provided for containers of MDA within the workplace. The labels shall comply with the requirements of 29 CFR 1910.1200(f).

- For pure MDA, the labels shall look similar to the sample at middle right.

- For mixtures containing MDA, the labels shall look similar to the sample at lower right.

- Employers shall obtain or develop, and shall provide access to a material safety data sheet (MSDS) for MDA. In meeting this obligation, employers shall make appropriate use of the information found in Appendices A and B of this text.

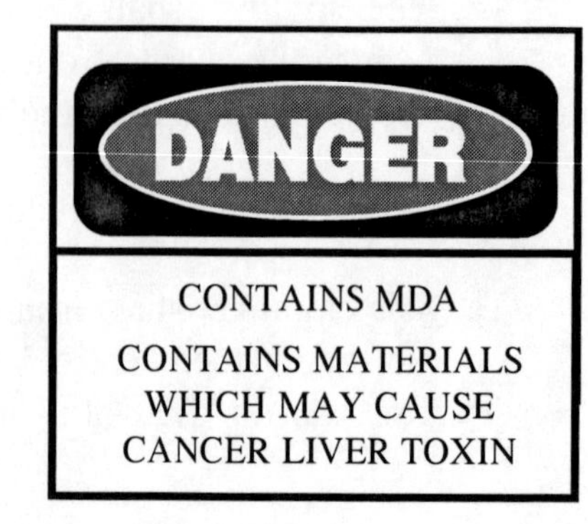

- Employers who are manufacturers or importers shall comply with the requirement in OSHA's Hazard Communication Standard, 29 CFR 1910.1200, to deliver an MSDS for MDA to downstream employers.

- The employer shall provide employees with information and training on MDA, in accordance with 29 CFR 1910.1200(h), at the time of initial assignment and at least annually thereafter.

- In addition to the information required under 29 CFR 1910.1200, the employer shall:
  - Provide an explanation of the contents of 29 CFR 1910.1050, including Appendices A and B, and indicate to employees where a copy of the standard is available;
  - Describe the medical surveillance program, and explain the information contained in 29 CFR 1910.1050, Appendix C; and
  - Describe the medical removal provisions.

- The employer shall make readily available, without cost, to all affected employees, all written materials relating to the employee training program, including a copy of 29 CFR 1910.1050.

- The employer shall provide, upon request, to the Assistant Secretary and the Director all information and training materials relating to the employee information and training program.

## Housekeeping

- All surfaces shall be maintained as free as practicable of visible accumulations of MDA.

- The employer shall institute a program for detecting MDA leaks, spills, and discharges, including regular visual inspections of operations involving liquid or solid MDA.

- All leaks shall be repaired and liquid or dust spills cleaned up promptly.

- Surfaces contaminated with MDA may not be cleaned by the use of compressed air.

- Shoveling, dry sweeping, and other methods of dry cleanup of MDA may be used where HEPA-filtered vacuuming and/or wet cleaning are not feasible or practical.

- Waste, scrap, debris, bags, containers, equipment, and clothing contaminated with MDA shall be collected and disposed of in a manner to prevent the reentry of MDA into the workplace.

## Medical Surveillance

- The employer shall make available a medical surveillance program for:
  - Employees exposed at or above the action level for 30 or more days per year;
  - Employees who are subject to dermal exposure to MDA for 15 or more days per year;
  - Employees who have been exposed in an emergency situation;
  - Employees whom the employer has reason to believe are being dermally exposed; and
  - Employees who show signs or symptoms of MDA exposure.

- ❑ The employer shall ensure that all medical examinations and procedures are performed, at a reasonable time and place, by, or under the supervision of, a licensed physician and provided without cost to the employee.

- ❑ Within 150 days of the effective date of this standard, or before the time of initial assignment, the employer shall provide each employee with a medical examination which includes:

  - A detailed history which includes:
    - Past work exposure to MDA or any other toxic substances;
    - A history of drugs, alcohol, tobacco, and medication routinely taken (duration and quantity); and
    - A history of dermatitis, chemical skin sensitization, or previous hepatic disease.
  - A physical examination which includes all routine physical examination parameters, skin examination, and signs of liver disease.
  - Laboratory tests, including liver function tests and urinalysis.
  - Additional tests as necessary in the opinion of the physician.

- ❑ No initial medical examination is required if adequate records show that the employee has been examined within the previous six months prior to the effective date of this standard or prior to the date of initial assignment.

- ❑ The employer shall provide each employee with a medical examination at least annually following the initial examination. These periodic examinations shall include at least the following elements:

  - A brief history regarding any new exposure to potential liver toxins, changes in drug, tobacco, and alcohol intake, and the appearance of physical signs relating to the liver and the skin;
  - The appropriate tests and examinations, including liver function tests and skin examinations; and
  - Appropriate additional tests or examinations as deemed necessary by the physician.

- ❑ If, in the physicians' opinion, the results of liver function tests indicate an abnormality, the employee shall be removed from further MDA exposure. Repeated liver function tests shall be conducted on advice of the physician.

- ❑ If the employer determines that the employee has been exposed to a potentially hazardous amount of MDA in an emergency situation, the employer shall provide medical examinations. If the results of liver function testing indicate an abnormality, the employee shall be removed. Repeated liver function tests shall be conducted on the advice of the physician. If the results of the tests are normal, tests must be repeated two to three weeks from the initial testing. No additional testing is required, if the physician so advises and the results of the second set of tests are normal.

- ❑ Where the employee develops signs and symptoms associated with exposure to MDA, the employer shall provide the employee with an additional medical examination, including a liver function test. Repeated liver function tests shall be conducted on the advice of the physician. If the results of the tests are normal, tests must be repeated two to three weeks from the initial testing. No additional testing is required, if the physician so advises and the results of the second set of tests are normal.

- If the employer selects the initial physician who conducts any medical examination or consultation provided to an employee under this section, and the employee has signs or symptoms of occupational exposure to MDA (which could include an abnormal liver function test), the employee disagrees with the opinion of the examining physician, and this opinion could affect the employee's job status, then the employee may designate an appropriate, mutually acceptable second physician to review any findings, determinations, or recommendations of the initial physician, and to conduct such examinations, consultations, and laboratory tests as the second physician deems necessary to facilitate this review.

- The employer shall promptly notify an employee of the right to seek a second medical opinion after each occasion that an initial physician conducts a medical examination or consultation. The employer may condition its participation in, and payment for, the multiple physician review mechanism upon the employee doing the following within 15 days after receipt of the foregoing notification or receipt of the initial physician's written opinion, whichever is later:
  - Informing the employer that he/she intends to seek a second medical opinion, and
  - Initiating steps to make an appointment with a second physician.

- If the findings, determinations, or recommendations of the second physician differ from those of the initial physician, then the employer and the employee shall assure that efforts are made for the two physicians to resolve any disagreement.

- If the two physicians have been unable to quickly resolve their disagreement, the employer and the employee through their respective physicians shall designate a third physician:
  - To review any findings, determinations, or recommendations of the prior physicians; and
  - To conduct such examinations, consultations, laboratory tests, and discussions with the prior physicians as the third physician deems necessary to resolve the disagreement of the prior physicians.

- The employer shall act consistently with the findings, determinations, and recommendations of the third physician, unless the employer and the employee reach an agreement that is otherwise consistent with the recommendations of at least one of the three physicians.

- The employer shall provide the following information to the examining physician:
  - A copy of 29 CFR 1910.1050 and its appendices;
  - A description of the affected employee's duties as they relate to the employee's potential exposure to MDA;
  - The employee's current actual or representative MDA exposure level;
  - A description of any personal protective equipment used or to be used; and
  - Information from previous employment-related medical examinations of the affected employee.

- ❑ The employer shall provide the foregoing information to a second physician under this section upon request either by the second physician or the employee.

- ❑ For each examination under this section, the employer shall obtain, and provide the employee with a copy of, the examining physician's written opinion within 15 days of its receipt. The written opinion shall include the following:

    - The occupationally pertinent results of the medical examination and tests;
    - The physician's opinion concerning whether the employee has any detected medical conditions which would place the employee at increased risk of material impairment of health from exposure to MDA;
    - The physician's recommended limitations upon the employee's exposure to MDA or upon the employee's use of protective clothing or equipment and respirators; and
    - A statement that the employee has been informed by the physician of the results of the medical examination and any medical conditions resulting from MDA exposure which require further explanation or treatment.

- ❑ The written opinion obtained by the employer shall not reveal specific findings or diagnoses unrelated to occupational exposures.

- ❑ The employee shall be removed from work environments in which exposure to MDA is at or above the action level or where dermal exposure to MDA may occur, following an initial examination, periodic examinations, an emergency situation, or an additional examination in the following circumstances:

    - The employee exhibits signs and/or symptoms indicative of acute exposure to MDA; or
    - The examining physician determines that an employee's abnormal liver function tests are not associated with MDA exposure but that the abnormalities may be exacerbated as a result of occupational exposure to MDA.

- ❑ The employer shall remove an employee from work environments in which exposure to MDA is at or above the action level, or where dermal exposure to MDA may occur, on each occasion that there is a final medical determination or opinion that the employee has a detected medical condition which places the employee at increased risk of material impairment to health from exposure to MDA.

- ❑ The phrase "final medical determination" shall mean the outcome of the physician review mechanism used pursuant to the medical surveillance provisions of this section.

- ❑ Where a final medical determination results in any recommended special protective measures for an employee, or limitations on an employee's exposure to MDA, the employer shall implement and act consistently with the recommendation.

- ❑ The employer shall return an employee to his or her former job status when:

    - The employee no longer shows signs or symptoms of exposure to MDA, upon the advice of the physician; or

- A subsequent final medical determination results in a medical finding, determination, or opinion that the employee no longer has a detected medical condition which places the employee at increased risk of material impairment to health from exposure to MDA.

- The requirement that an employer return an employee to his or her former job status is not intended to expand upon or restrict any rights an employee has or would have had, absent temporary medical removal, to a specific job classification or position under the terms of a collective bargaining agreement.

- The employer shall remove any limitations placed on an employee, or end any special protective measures provided to an employee, pursuant to a final medical determination, when a subsequent final medical determination indicates that the limitations or special protective measures are no longer necessary.

- Where the physician review mechanism used pursuant to the medical surveillance provisions, has not yet resulted in a final medical determination with respect to an employee, the employer shall act as follows:
  - The employer may remove the employee from exposure to MDA, provide special protective measures to the employee, or place limitations upon the employee, consistent with the medical findings, determinations, or recommendations of any of the physicians who have reviewed the employee's health status.
  - The employer may return the employee to his or her former job status and end any special protective measures provided to the employee, consistent with the medical findings, determinations, or recommendations of any of the physicians who have reviewed the employee's health status, with two exceptions:
    - If the initial removal, special protection, or limitation of the employee resulted from a final medical determination which differed from the findings, determinations, or recommendations of the initial physician; or
    - If the employee has been on removal status for the preceding six months as a result of exposure to MDA.

- The employer shall provide to an employee up to six months of medical removal protection benefits on each occasion that an employee is removed from exposure to MDA.

- The requirement that an employer provide medical removal protection benefits means that the employer shall maintain the wage rate, seniority, and other employment rights and benefits of the employee as though the employee had not been removed from normal exposure to MDA or otherwise limited.

- During the period of time that an employee is removed from normal exposure to MDA or otherwise limited, the employer may condition the provision of medical removal protection benefits upon the employee's participation in follow-up medical surveillance made available.

- If a removed employee files a claim for workers' compensation payments for an MDA-related disability, then the employer shall continue to provide medical removal protection benefits pending disposition of the claim. To the extent that an award is made to the employee for earnings lost during the period of removal, the employer's medical removal protection obligation shall be reduced by such amount. The employer shall receive no credit for workers' compensation payments received by the employee for treatment-related expenses.

- ❑ The employer's obligation to provide medical removal protection benefits to a removed employee shall be reduced to the extent that the employee receives compensation for earnings lost during the period of removal either from a publicly or employer-funded compensation program, or receives income from non-MDA-related employment with any employer made possible by virtue of the employee's removal.

- ❑ The employer shall take the following measures with respect to any employee removed from exposure to MDA employees who do not recover within the six months of removal.
    - The employer shall make available to the employee a medical examination to obtain a final medical determination with respect to the employee;
    - The employer shall assure that the final medical determination obtained indicates whether or not the employee may be returned to his or her former job status, and, if not, what steps should be taken to protect the employee's health;
    - Where the final medical determination has not yet been obtained, or, once obtained, indicates that the employee may not yet be returned to his or her former job status, the employer shall continue to provide medical removal protection benefits to the employee until either the employee is returned to former job status, or a final medical determination is made that the employee is incapable of ever safely returning to his or her former job status; and
    - Where the employer acts pursuant to a final medical determination which permits the return of the employee to his or her former job status, despite what would otherwise be an abnormal liver function test, later questions concerning removing the employee again shall be decided by a final medical determination. The employer need not automatically remove such an employee.

- ❑ Where an employer removes an employee from exposure to MDA or otherwise places limitations on an employee due to the effects of MDA exposure on the employee's medical condition, the employer shall provide medical removal protection benefits to the employee.

## Recordkeeping

- ❑ Where, as a result of the initial monitoring, the processing, use, or handling of products made from or containing MDA are exempted, the employer shall establish and maintain an accurate record of monitoring relied on in support of the exemption.

- ❑ This record shall include at least the following information:
    - The product qualifying for exemption;
    - The source of the monitoring data (e.g., performed by the employer or a private contractor);
    - The testing protocol, results of testing, and/or analysis of the material for the release of MDA;

- A description of the operation exempted and how the data support the exemption (e.g., a description of the way the monitoring data representative of the conditions at the affected facility); and
- Other data relevant to the operations, materials, processing, or employee exposures covered by the exemption.

❑ The employer shall maintain this record for the duration of the employer's reliance upon such monitoring data.

❑ Where the processing, use, or handling of products made from or containing MDA are exempted, the employer shall establish and maintain an accurate record of objective data relied upon in support of the exemption.

❑ This record shall include at least the following information:

- The product qualifying for exemption;
- The source of the objective data;
- The testing protocol, results of testing, and/or analysis of the material for the release of MDA;
- A description of the operation exempted and how the data support the exemption; and
- Other data relevant to the operations, materials, processing, or employee exposures covered by the exemption.

❑ The employer shall maintain this record for the duration of the employer's reliance upon such objective data.

❑ The employer shall establish and maintain an accurate record of all measurements, in accordance with 29 CFR 1910.20.

❑ This record shall include:

- The dates, number, duration, and results of each of the samples taken, including a description of the procedure used to determine representative employee exposures;
- Identification of the sampling and analytical methods used;
- A description of the type of respiratory protective devices worn, if any; and
- The name, social security number, job classification, and exposure levels of the employee monitored and all other employees whose exposure the measurement is intended to represent.

❑ The employer shall maintain this record for at least 30 years, in accordance with 29 CFR 1910.20.

❑ The employer shall establish and maintain an accurate record for each employee subject to medical surveillance, in accordance with 29 CFR 1910.20.

❑ This record shall include:

- The name, social security number, and description of the duties of the employee;
- The employer's copy of the physician's written opinion on initial, periodic, and any special examinations, including results of medical examination and all tests, opinions, and recommendations;
- Results of any airborne exposure monitoring done for that employee and the representative exposure levels supplied to the physician; and
- Any employee medical complaints related to exposure to MDA.

❑ The employer shall keep, or assure that the examining physician keeps, the following medical records:

- A copy of 29 CFR 1910.1050 and its appendices, except that the employer may keep one copy of the standard and its appendices for all employees provided the employer references the standard and its appendices in the medical surveillance record of each employee;
- A copy of the information provided to the physician;
- A description of the laboratory procedures and a copy of any standards or guidelines used to interpret the test results or references to the information; and
- A copy of the employee's medical and work history related to exposure to MDA.

❑ The employer shall maintain this record for at least the duration of employment plus 30 years, in accordance with 29 CFR 1910.20.

❑ The employer shall establish and maintain an accurate record for each employee removed from current exposure to MDA.

❑ Each record shall include:

- The name and social security number of the employee;
- The date of each occasion that the employee was removed from current exposure to MDA as well as the corresponding date on which the employee was returned to his or her former job status;
- A brief explanation of how each removal was or is being accomplished; and
- A statement with respect to each removal indicating the reason for the removal.

❑ The employer shall maintain each medical removal record for at least the duration of an employee's employment plus 30 years.

❑ The employer shall assure that records required to be maintained by this section shall be made available, upon request, to the Assistant Secretary and the Director for examination and copying.

❑ Employee exposure monitoring records required by this section shall be provided, upon request, for examination and copying to employees, employee representatives, and the Assistant Secretary, in accordance with 29 CFR 1910.20(a) through (e) and (g) through (i).

- Employee medical records required by this section shall be provided, upon request, for examination and copying, to the subject employee, to anyone having the specific written consent of the subject employee, and to the Assistant Secretary, in accordance with 29 CFR 1910.20.

- The employer shall comply with the requirements involving transfer of records set forth in 29 CFR 1910.20(h).

- If the employer ceases to do business and there is no successor employer to receive and retain the records for the prescribed period, the employer shall notify the Director, at least 90 days prior to disposal, and transmit the records to the Director if so requested by the Director within that period.

## Observation of Monitoring

- The employer shall provide affected employees, or their designated representatives, an opportunity to observe the measuring or monitoring of employee exposure to MDA.

- When observation of the measuring or monitoring of employee exposure to MDA requires entry into areas where the use of protective clothing and equipment or respirators is required, the employer shall provide the observer with personal protective clothing and equipment or respirators required to be worn by employees working in the area, assure the use of such clothing and equipment or respirators, and require the observer to comply with all other applicable safety and health procedures.

# Vinyl Chloride

## 1910.1017

Section 16 applies to the manufacture, reaction, packaging, repackaging, storage, handling, or use of vinyl chloride or polyvinyl chloride, but does not apply to the handling or use of fabricated products made of polyvinyl chloride. This section also applies to the transportation of vinyl chloride or polyvinyl chloride except to the extent that the Department of Transportation may regulate the hazards covered by this section.

### Definitions

- *Action level* means a concentration of vinyl chloride of 0.5 parts per million (0.5 ppm) averaged over an 8-hour workday.
- *Assistant Secretary* means the Assistant Secretary of Labor for Occupational Safety and Health, U.S. Department of Labor, or his designee.
- *Authorized person* means any person specifically authorized by the employer whose duties require him to enter a regulated area or any person entering such an area as a designated representative of employees for the purpose of exercising an opportunity to observe monitoring and measuring procedures.
- *Director* means the Director, National Institute for Occupational Safety and Health, U.S. Department of Health and Human Services, or his designee.
- *Emergency* means any occurrence such as, but not limited to, equipment failure or operation of a relief device which is likely to, or does, result in massive release of vinyl chloride.
- *Fabricated product* means a product made wholly or partly from polyvinyl chloride, and which does not require further processing at temperatures or times sufficient to cause mass melting of the polyvinyl chloride resulting in the release of vinyl chloride.

- *Hazardous operation* means any operation, procedure, or activity where a release of either vinyl chloride liquid or gas might be expected as a consequence of the operation, or accident in the operation, which would result in an employee exposure in excess of the permissible exposure limit.

- *OSHA Area Director* means the Director for the Occupational Safety and Health Administration Area Office having jurisdiction over the geographic area in which the employer's establishment is located.

- *Polyvinyl chloride* means polyvinyl chloride homopolymer or copolymer before such is converted to a fabricated product.

- *Vinyl chloride* means vinyl chloride monomer.

## Permissible Exposure Limit (PEL)

- No employee may be exposed to vinyl chloride at concentrations greater than 1 part per million (ppm) averaged over any 8-hour period.

- No employee may be exposed to vinyl chloride at concentrations greater than 5 ppm averaged over any period not exceeding 15 minutes.

- No employee may be exposed to vinyl chloride by direct contact with liquid vinyl chloride.

## Exposure Monitoring

- A program of initial monitoring and measurement shall be undertaken in each establishment to determine if there is any employee exposed, without regard to the use of respirators, in excess of the action level.

- Where a determination shows any employee exposures, without regard to the use of respirators, in excess of the action level, a program for determining exposures for each such employee shall be established. Such a program:

  - Shall be repeated at least monthly where any employee is exposed, without regard to the use of respirators, in excess of the PEL.
  - Shall be repeated not less than quarterly where any employee is exposed, without regard to the use of respirators, in excess of the action level.
  - May be discontinued for any employee only when at least two consecutive monitoring determinations, made not less than five working days apart, show exposures for that employee at or below the action level.

- Whenever there has been a production, process, or control change which may result in an increase in the release of vinyl chloride, or the employer has any other reason to suspect that any employee may be exposed in excess of the action level, a determination of employee exposure shall be performed.

- The method of monitoring and measurement shall have an accuracy, to a confidence level of 95 percent, of not less than plus or minus 50 percent from 0.25 through 0.5 ppm; plus or minus 35 percent from over 0.5 ppm through 1.0 ppm; and plus or minus 25 percent over 1.0 ppm. (Methods meeting these accuracy requirements are available in the "NIOSH Manual of Analytical Methods.")

- Employees, or their designated representatives, shall be afforded reasonable opportunity to observe the monitoring and measuring.

## Regulated Areas

- A regulated area shall be established where:
  - Vinyl chloride or polyvinyl chloride is manufactured, reacted, repackaged, stored, handled, or used; and
  - Vinyl chloride concentrations are in excess of the PEL.

- Access to regulated areas shall be limited to authorized persons.

## Methods of Compliance

- Employee exposures to vinyl chloride shall be controlled to at or below the PEL provided by engineering, work-practice, and personal protective controls feasible engineering and work-practice controls shall immediately be used to reduce exposures to at or below the PEL.

- Wherever feasible engineering and work-practice controls which can be instituted immediately are not sufficient to reduce exposures to at or below the PEL, they shall nonetheless be used to reduce exposures to the lowest practicable level, and shall be supplemented by respiratory protection. A program shall be established and implemented to reduce exposures to at or below the PEL, or to the greatest extent feasible, solely by means of engineering and work-practice controls, as soon as feasible.

- Written plans for such a program shall be developed and furnished, upon request, to authorized representatives of the Assistant Secretary and the Director, for examination and copying. Such plans shall be updated at least every six months.

## Respiratory Protection

- For employees who use respirators, the employer must provide respirators that comply with the requirements of this paragraph.

- The employer must implement a respiratory protection program in accordance with 29 CFR 1910.134(b) through (d) (except (d)(1)(iii) and (d)(3)(iii)(B)(1) and (2)) and (f) through (m).

- Respirators must be selected from Table 4-16-1.

## Table 4-16-1: Respiratory Protection for Vinyl Chloride

| Atmospheric concentration of vinyl chloride | Required respirator |
|---|---|
| Unknown or above 3,600 p/m | Open-circuit, self-contained breathing apparatus, pressure-demand type, with full facepiece. |
| Not over 3,600 p/m | Combination Type C supplied-air respirator, pressure-demand type, with full or half facepiece, and auxiliary self-contained air supply; or |
| | Combination type, supplied-air respirator, continuous flow type, with full or half facepiece, and auxiliary self-contained air supply. |
| Not over 1,000 p/m | Type C supplied air respirator, continuous flow type, with full or half facepiece, helmet, or hood. |
| Not over 100 p/m | Combination Type C supplied-air respirator demand type, with full facepiece, and auxiliary self-contained air supply; or |
| | Open-circuit self-contained breathing apparatus with full facepiece, in demand mode; or |
| | Type C supplied-air respirator, demand type, with full facepiece. |
| Not over 25 p/m | A powered air-purifying respirator with hood, helmet, full or half facepiece, and a canister which provides a service life of at least 4 hours for concentrations of vinyl chloride up to 25 p/m; or |
| | Gas mask, front- or back-mounted canister which provides a service life of at least 4 hours for concentrations of vinyl chloride up to 25 p/m. |
| Not over 10 p/m | Combination Type C supplied-air respirator, demand type, with half facepiece, and auxiliary self-contained air supply; or |
| | Type C supplied-air respirator, demand type, with half facepiece; or |
| | Any chemical cartridge respirator with an organic vapor cartridge which provides a service life of at least 1 hour for concentrations of vinyl chloride up to 10 p/m. |

Note: Respirators specified for higher concentrations may be used for lower concentrations.

- ❑ When air-purifying respirators are used:
  - Air-purifying canisters or cartridges must be replaced prior to the expiration of their service life or at the end of the shift in which they are first used, whichever occurs first.
  - A continuous-monitoring and alarm system must be provided when concentrations of vinyl chloride could reasonably exceed the allowable concentrations for the devices in use. Such a system must be used to alert employees when vinyl chloride concentrations exceed the allowable concentrations for the devices in use.

## Hazardous Operations

- ❑ Employees engaged in hazardous operations, including entry of vessels to clean polyvinyl chloride residue from vessel walls, shall be provided with and required to wear and use respiratory protection and protective garments to prevent skin contact with liquid vinyl chloride or with polyvinyl chloride residue from vessel walls.

- The protective garments shall be selected for the operation and its possible exposure conditions.
- Clean and dry protective garments shall be provided for each use.

## Emergency Situations

- A written operational plan for emergency situations shall be developed for each facility storing, handling, or otherwise using vinyl chloride as a liquid or compressed gas. Appropriate portions of the plan shall be implemented in the event of an emergency. The plan shall specifically provide that:
    - Employees engaged in hazardous operations or correcting situations of existing hazardous releases shall be properly equipped; and
    - Other employees not so equipped shall evacuate the area and not return until conditions are controlled and the emergency is abated.

## Medical Surveillance

- A program of medical surveillance shall be instituted for each employee exposed to vinyl chloride in excess of the action level, without regard to the use of respirators. The program shall provide each such employee with an opportunity for examinations and tests. All medical examinations and procedures shall be performed by or under the supervision of a licensed physician, and shall be provided without cost to the employee at the time of initial assignment, or upon institution of medical surveillance.
- A general physical examination shall be performed, with specific attention to detecting enlargement of the liver, spleen, kidneys, or dysfunction in these organs; and for abnormalities in the skin, connective tissues, and the pulmonary system.
- A medical history shall be taken, which shall include the following topics:
    - Alcohol intake;
    - Past history of hepatitis;
    - Work history and past exposure to potential hepatotoxic agents, including drugs and chemicals;
    - Past history of blood transfusions;
    - Past history of hospitalizations; and
    - A serum specimen shall be obtained and determinations made of:
        - Total bilirubin;
        - Alkaline phosphatase;

    - Serum glutamic oxalacetic transaminase (SGOT);
    - Serum glutamic pyruvic transaminase (SGPT); and
    - Gamma glustamyl transpeptidase.

- ❑ Examinations shall be performed at least:
  - Every six months for each employee who has been employed in vinyl chloride or polyvinyl chloride manufacturing for ten years or longer; and
  - Annually for all other employees.
- ❑ Each employee exposed in an emergency shall be afforded appropriate medical surveillance.
- ❑ A statement of each employee's suitability for continued exposure to vinyl chloride, including use of protective equipment and respirators, shall be obtained from the examining physician promptly after any examination. A copy of the physician's statement shall be provided to each employee.
- ❑ If any employee's health would be materially impaired by continued exposure, such employee shall be withdrawn from possible contact with vinyl chloride.
- ❑ Laboratory analyses for all biological specimens included in medical examinations shall be performed in laboratories licensed under 42 CFR Part 74.
- ❑ If the examining physician determines that alternative medical examinations will provide at least equal assurance of detecting medical conditions pertinent to the exposure to vinyl chloride, the employer may accept such alternative examinations. The employer must obtain a statement from the examining physician setting forth the alternative examinations and the rationale for substitution. This statement shall be available, upon request, to authorized representatives of the Assistant Secretary and the Director, for examination and copying.

## Communication of Hazard to Employees

- ❑ Each employee engaged in vinyl chloride or polyvinyl chloride operations shall be provided training in a program relating to the hazards of vinyl chloride and precautions for its safe use.
- ❑ The program shall include:
  - The nature of the health hazard from chronic exposure to vinyl chloride, including specifically the carcinogenic hazard;
  - The specific nature of operations which could result in exposure to vinyl chloride in excess of the permissible limit and necessary protective steps;
  - The purpose, proper use, and limitations of respiratory protective devices;
  - The fire hazard and acute toxicity of vinyl chloride, and the necessary protective steps;
  - The purpose and description of, the monitoring program;
  - The purpose and description of, the medical surveillance program;
  - Emergency procedures;

- Specific information to aid the employee in recognition of conditions which may result in the release of vinyl chloride; and
- A review of 29 CFR 1910.1017 at the employee's first training and indoctrination program, and annually thereafter.

- All materials relating to the program shall be provided, upon request, to the Assistant Secretary and the Director.
- Entrances to regulated areas shall be posted with legible signs. (See sample #1)
- Areas containing hazardous operations, or where an emergency currently exists, shall be posted with legible signs. (See sample #2)
- Containers of polyvinyl chloride resin waste from reactors or other waste contaminated with vinyl chloride shall be legibly labeled. (See sample #3)
- Containers of polyvinyl chloride shall be legibly labeled. (See sample #4)
- Containers of vinyl chloride shall be legibly labeled applied near the label or placard. (See sample #5)
- No statement shall appear on or near any required sign, label, or instruction that contradicts or detracts from the effect of any required warning, information, or instruction.

## Recordkeeping

- All records shall include the name and social security number of each employee, where relevant.
- Records of required monitoring and measuring and medical records shall be provided, upon request, to employees, designated employee representatives, and the Assistant Secretary in accordance with 29 CFR 1910.20(a) through (e), and (g) through (i). These records shall be provided, upon request, to the Director. Authorized personnel rosters shall also be provided, upon request, to the Assistant Secretary and the Director.

1

Cancer-Suspect Agent Area

Authorized Personnel Only

2

Cancer-Suspect Agent
Protective Equipment
Required

Authorized Personnel Only

3

Contaminated With
Vinyl Chloride

Cancer-Suspect Agent

4

Polyvinyl Chloride
(or Trade Name)

Contains Vinyl Chloride

Vinyl Chloride is a
Cancer-Suspect Agent

5

Vinyl Chloride

Extremely Flammable
Gas Under Pressure

Cancer Suspect Agent

- ❑ Monitoring and measuring records shall:
  - State the date of such monitoring and measuring, the concentrations determined, and identify the instruments and methods used; and
  - Include any additional information necessary to determine individual employee exposures where such exposures are determined by means other than individual monitoring of employees.
- ❑ Medical records shall be maintained for 30 years, or the duration of the employment of each employee plus 20 years, whichever is longer.
- ❑ In the event that the employer ceases to do business and there is no successor to receive and retain his records for the prescribed period, these records shall be transmitted by registered mail to the Director, and each employee individually notified in writing of this transfer. The employer shall also comply with any additional requirements set forth in 29 CFR 1910.20(h).

## Reports

- ❑ Not later than one month after the establishment of a regulated area, the following information shall be reported to the OSHA Area Director and any changes to such information shall be reported within 15 days:
  - The address and location of each establishment which has one or more regulated areas; and
  - The number of employees in each regulated area during normal operations, including maintenance.
- ❑ Emergencies, and the facts obtainable at that time, shall be reported within 24 hours to the OSHA Area Director. Upon request of the Area Director, the employer shall submit additional information in writing relevant to the nature and extent of employee exposures and measures taken to prevent future emergencies of similar nature.
- ❑ Within ten working days following any monitoring and measuring which discloses that any employee has been exposed in excess of the PEL, without regard to the use of respirators, each such employee shall be notified in writing of the results of the exposure measurement and the steps being taken to reduce the exposure to within the PEL.

# Appendix A

# Sample Forms

The following pages contain forms that may be copied and used for your program. You may cutout or copy these forms or customize your own with more or less information. Included forms are:

1. Workplace Chemical List
2. Chemical Inventory List
3. MSDS Request Form
4. Training Certification Form

## Workplace Chemical List

| MSDS # | Description | MSDS Date | Hazard Rating | PPE | Flash-point | First Aid | Spill | Disposal | Location |
|---|---|---|---|---|---|---|---|---|---|
| | | | Health -<br>Fire -<br>Reactivity -<br>Specific - | | | Eyes -<br>Skin -<br>Inhale -<br>Ingest - | | | |
| | | | Health -<br>Fire -<br>Reactivity -<br>Specific - | | | Eyes -<br>Skin -<br>Inhale -<br>Ingest - | | | |
| | | | Health -<br>Fire -<br>Reactivity -<br>Specific - | | | Eyes -<br>Skin -<br>Inhale -<br>Ingest - | | | |
| | | | Health -<br>Fire -<br>Reactivity -<br>Specific - | | | Eyes -<br>Skin -<br>Inhale -<br>Ingest - | | | |
| | | | Health -<br>Fire -<br>Reactivity -<br>Specific - | | | Eyes -<br>Skin -<br>Inhale -<br>Ingest - | | | |
| | | | Health -<br>Fire -<br>Reactivity -<br>Specific - | | | Eyes -<br>Skin -<br>Inhale -<br>Ingest - | | | |
| | | | Health -<br>Fire -<br>Reactivity -<br>Specific - | | | Eyes -<br>Skin -<br>Inhale -<br>Ingest - | | | |
| | | | Health -<br>Fire -<br>Reactivity -<br>Specific - | | | Eyes -<br>Skin -<br>Inhale -<br>Ingest - | | | |
| | | | Health -<br>Fire -<br>Reactivity -<br>Specific - | | | Eyes -<br>Skin -<br>Inhale -<br>Ingest - | | | |
| | | | Health -<br>Fire -<br>Reactivity -<br>Specific - | | | Eyes -<br>Skin -<br>Inhale -<br>Ingest - | | | |

## Chemical Inventory List

This list has been issued to the supervisor of each shop and/or department to determine what chemical products are used at each location. Please evaluate each container, including all drums, cans, bottles, pipes, etc., to determine the contents and any other available information. Include material safety and informational data sheets for each of the products you find. This information will be used to develop an accurate workplace chemical list as required by the Hazard Communication Standard.

| Product Name/Alternate Name | Manufacturer | Type of Product (Adhesive, Paint, etc.) |
|---|---|---|
| 1. ____________ | ____________ | ____________ |
| 2. ____________ | ____________ | ____________ |
| 3. ____________ | ____________ | ____________ |
| 4. ____________ | ____________ | ____________ |
| 5. ____________ | ____________ | ____________ |
| 6. ____________ | ____________ | ____________ |
| 7. ____________ | ____________ | ____________ |
| 8. ____________ | ____________ | ____________ |
| 9. ____________ | ____________ | ____________ |
| 10. ____________ | ____________ | ____________ |
| 11. ____________ | ____________ | ____________ |
| 12. ____________ | ____________ | ____________ |
| 13. ____________ | ____________ | ____________ |
| 14. ____________ | ____________ | ____________ |
| 15. ____________ | ____________ | ____________ |
| 16. ____________ | ____________ | ____________ |

Make additional copies and attach, if necessary. Page ___ of ___

## Material Safety Data Sheet Request Form

Dear ___________________:

The Occupational Safety and Health Administration (OSHA) Hazard Communication Standard, Superfund Amendments (SARA Title III), and company policy require us to maintain and distribute material safety data sheets (MSDSs) for all chemicals, substances, and hazardous materials used in our facility. It also requires us to make these MSDSs available to employees who utilize, or may potentially be exposed to, these hazardous chemicals, materials, and/or substances.

In order to meet these regulations and policies, we request a completed MSDS (OSHA Form 174 or equivalent) for the following product(s):

Product Name

_______________________________________

_______________________________________

_______________________________________

_______________________________________

_______________________________________

We also request any additional information, supplemental MSDSs, or any relevant data your company or supplier currently has, or may acquire in the future, concerning the safety and health aspects of the product(s).

Please remit this information to the following address:

## Training Certification Form

Employee Name ________________________________________

On the date signed below, hazard communication and chemical safety training was provided. Training consisted of:

1. Hazard Communication Requirements

   A. HazCom terms and company policy statement

   B. Labeling of containers

   C. Understanding and use of material safety data sheets (MSDSs)

   D. Understanding and use of the workplace chemical list and guide indices

   E. Safe handling and storage of chemical products

2. Hazard Identification

3. Location of written Hazard Communication Program and MSDSs

4. Action to take on a spill or fire involving chemical products

5. Content of MSDSs

6. Use of personal protective equipment

7. Health hazard characterization

8. Workplace chemicals and department/shop-specific instruction

9. Specific responsibilities of employee

### Employee Verification & Certification

I certify that I have received the above training on chemical product safety and am aware of my responsibilities for safe chemical use, storage, handling, and emergency procedures.

Employee Signature: ________________________ Date __________

Trainer Signature: __________________________ Date __________

# Appendix B

# OSHA Regional Offices

In case of emergency, call 1-800-321-OSHA.

**Region 1**
**CT, MA, ME, NH, RI, VT**
JFK Federal Building, Room E340
Boston, MA 02203
Telephone: (617) 565-7164
Fax: (617) 565-9827

**Region 2**
**NJ, NY, Puerto Rico, Virgin Islands**
201 Varick Street, Room 670
New York, NY 10014
Telephone: (212) 337-2378
Fax: (212) 337-2371

**Region 3**
**DC, DE, MD, PA, VA, WV**
Gateway Building, Suite 2100
3535 Market Street
Philadelphia, PA 19104
Telephone: (215) 596-1201
Fax: (215) 596-4872

**Region 4**
**AL, FL, GA, KY, MS, NC, SC, TN**
61 Forsyth Street, SW
Atlanta, GA 30303
Telephone: (404) 562-2300
Fax: (404) 562-2295

**Region 5**
**IL, IN, MI, MN, OH, WI**
230 South Dearborn Street, Room 3244
Chicago, IL 60604
Telephone: (312) 353-2220
Fax: (312) 353-7774

**Region 6**
**AR, LA, NM, OK, TX**
525 Griffin Street, Room 602
Dallas, Texas 75202
Telephone: (214) 767-4731
Fax: (214) 767-4137

**Region 7**
**IA, KS, MO, NE**
1100 Main Street, Suite 800
Kansas City, MO 64105
Telephone: (816) 426-5861
Fax: (816) 426-2750

**Region 8**
**CO, MT, ND, SD, UT, WY**
1999 Broadway, Suite 1690
Denver, CO 80202-5716
Telephone: (303) 844-1600
Fax: (303) 844-1616

**Region 9**
**AZ, CA, Guam, HI, NV**
71 Stevenson Street, Room 420
San Francisco, CA 94105
Telephone: (415) 975-4310
Fax: (415) 975-4319

**Region 10**
**AK, ID, OR, WA**
1111 Third Avenue, Suite 715
Seattle, WA 98101-3212
Telephone: (206) 553-5930
Fax: (206) 553-6499

# Index

# Government Institutes Mini-Catalog

| PC # | | ENVIRONMENTAL TITLES | Pub Date | Price |
|---|---|---|---|---|
| 629 | | ABCs of Environmental Regulation: Understanding the Federal Regulations | 1998 | $49 |
| 627 | | ABCs of Environmental Science | 1998 | $39 |
| 672 | | Book of Lists for Regulated Hazardous Substances, 9th Edition | 1999 | $79 |
| 579 | | Brownfields Redevelopment | 1998 | $79 |
| 4100 | CD | CFR Chemical Lists on CD ROM, 1999 Edition | 1999 | $125 |
| 4089 | Disk | Chemical Data for Workplace Sampling & Analysis, Single User Disk | 1997 | $125 |
| 512 | | Clean Water Handbook, Second Edition | 1996 | $89 |
| 581 | | EH&S Auditing Made Easy | 1997 | $79 |
| 673 | | EH& S CFR Training Requirements, Fourth Edition | 1999 | $89 |
| 4082 | CD | EMMI-Envl Monitoring Methods Index for Windows-Network | 1997 | $537 |
| 4082 | CD | EMMI-Envl Monitoring Methods Index for Windows-Single User | 1997 | $179 |
| 525 | | Environmental Audits, 7th Edition | 1996 | $79 |
| 548 | | Environmental Engineering and Science: An Introduction | 1997 | $79 |
| 643 | | Environmental Guide to the Internet, Fourth Edition | 1998 | $59 |
| 650 | | Environmental Law Handbook, Fifteenth Edition | 1999 | $89 |
| 688 | | Environmental Health & Safety Dictionary, Seventh Edition | 2000 | $79 |
| 676 | | Environmental Statutes, 2000 Edition | 2000 | $89 |
| 4097 | CD | OSHA CFRs Made Easy (29 CFRs)/CD ROM | 1998 | $129 |
| 4102 | CD | 1999 Title 21 Food & Drug CFRs on CD ROM-Single User | 1999 | $325 |
| 4099 | CD | Environmental Statutes on CD ROM for Windows-Single User | 1999 | $139 |
| 570 | | Environmentalism at the Crossroads | 1995 | $39 |
| 536 | | ESAs Made Easy | 1996 | $59 |
| 515 | | Industrial Environmental Management: A Practical Approach | 1996 | $79 |
| 510 | | ISO 14000: Understanding Environmental Standards | 1996 | $69 |
| 551 | | ISO 14001: An Executive Report | 1996 | $55 |
| 588 | | International Environmental Auditing | 1998 | $149 |
| 518 | | Lead Regulation Handbook | 1996 | $79 |
| 554 | | Property Rights: Understanding Government Takings | 1997 | $79 |
| 582 | | Recycling & Waste Mgmt Guide to the Internet | 1997 | $49 |
| 615 | | Risk Management Planning Handbook | 1998 | $89 |
| 603 | | Superfund Manual, 6th Edition | 1997 | $115 |
| 566 | | TSCA Handbook, Third Edition | 1997 | $95 |
| 534 | | Wetland Mitigation: Mitigation Banking and Other Strategies | 1997 | $75 |

| PC # | SAFETY and HEALTH TITLES | Pub Date | Price |
|---|---|---|---|
| 547 | Construction Safety Handbook | 1996 | $79 |
| 553 | Cumulative Trauma Disorders | 1997 | $59 |
| 663 | Forklift Safety, Second Edition | 1999 | $69 |
| 539 | Fundamentals of Occupational Safety & Health | 1996 | $49 |
| 612 | HAZWOPER Incident Command | 1998 | $59 |
| 535 | Making Sense of OSHA Compliance | 1997 | $59 |
| 589 | Managing Fatigue in Transportation, *ATA Conference* | 1997 | $75 |
| 558 | PPE Made Easy | 1998 | $79 |
| 598 | Project Management for EH&S Professionals | 1997 | $59 |
| 552 | Safety & Health in Agriculture, Forestry and Fisheries | 1997 | $125 |
| 669 | Safety & Health on the Internet, Third Edition | 1999 | $59 |
| 597 | Safety Is A People Business | 1997 | $49 |
| 668 | Safety Made Easy, Second Edition | 1999 | $59 |
| 590 | Your Company Safety and Health Manual | 1997 | $79 |

# Government Institutes Order Form

4 Research Place, Suite 200 • Rockville, MD 20850-3226
Tel (301) 921-2323 • Fax (301) 921-0264
Internet: http://www.govinst.com • E-mail: giinfo@govinst.com

## 4 EASY WAYS TO ORDER

**1. Tel:** **(301) 921-2323**
Have your credit card ready when you call.

**2. Fax:** **(301) 921-0264**
Fax this completed order form with your company purchase order or credit card information.

**3. Mail:** **Government Institutes Division**
ABS Group Inc.
P.O. Box 846304
Dallas, TX 75284-6304 USA
Mail this completed order form with a check, company purchase order, or credit card information.

**4. Online:** **Visit http://www.govinst.com**

## PAYMENT OPTIONS

❑ **Check** *(payable in US dollars to* ***ABS Group Inc. Government Institutes Division****)*

❑ **Purchase Order** *(This order form must be attached to your company P.O. Note: All International orders must be prepaid.)*

❑ **Credit Card** ☐ VISA ☐ MasterCard ☐ American Express

Exp. ___ /___

Credit Card No. ____________________

Signature ____________________

(Government Institutes' Federal I.D.# is 13-2695912)

## CUSTOMER INFORMATION

**Ship To:** (Please attach your purchase order)

Name: ____________________

GI Account # (*7 digits on mailing label*): ____________

Company/Institution: ____________________

Address: ____________________
(Please supply street address for UPS shipping)

City: ____________ State/Province: ________

Zip/Postal Code: __________ Country: ________

Tel: ( )

Fax: ( )

Email Address: ____________________

**Bill To:** (if different from ship-to address)

Name: ____________________

Title/Position: ____________________

Company/Institution: ____________________

Address: ____________________
(Please supply street address for UPS shipping)

City: ____________ State/Province: ________

Zip/Postal Code: __________ Country: ________

Tel: ( )

Fax: ( )

Email Address: ____________________

| Qty. | Product Code | Title | Price |
|---|---|---|---|
| | | | |
| | | | |
| | | | |
| | | | |
| | | | |
| | | | |

❑ **New Edition No Obligation Standing Order Program**
Please enroll me in this program for the products I have ordered. Government Institutes will notify me of new editions by sending me an invoice. I understand that there is no obligation to purchase the product. This invoice is simply my reminder that a new edition has been released.

Subtotal ________
MD Residents add 5% Sales Tax ________
Shipping and Handling (see box below) ________
**Total Payment Enclosed** ________

**15 DAY MONEY-BACK GUARANTEE**
If you're not completely satisfied with any product, return it undamaged within 15 days for a full and immediate refund on the price of the product.

**SOURCE CODE: BP01**

| Shipping and Handling | Sales Tax |
|---|---|
| **Within U.S:** | Maryland .............. 5% |
| 1-4 products: $6/product | Tennessee .......... 6% |
| 5 or more: $4/product | Texas ............. 8.25% |
| **Outside U.S:** | Virginia ............. 4.5% |
| Add $15 for each item (Global) | |